BRAIDS, LINKS, AND MAPPING CLASS GROUPS

BY

JOAN S. BIRMAN

based on lecture notes by James Cannon

PRINCETON UNIVERSITY PRESS
PRINCETON, NEW JERSEY
1975

Published by Princeton University Press, 41 William Street,
Princeton, New Jersey 08540

LCC 74-2961

ISBN 0-691-08149-2

Printed in the United States of America

5 7 9 10 8 6 4

Library of Congress Cataloging-in-Publication data will be found on the
last printed page of this book

Annals of Mathematics Studies

Number 82

To the memory of Ralph H. Fox

PREFACE

This manuscript is based upon lectures given at Princeton University during the fall semester of 1971-72. The central theme is Artin's braid group, and the many ways that the notion of a braid has proved to be important in low dimensional topology.

Chapter 1 is concerned with the concept of a braid as a group of motions of points in a manifold. Structural and algebraic properties of the braid groups of two manifolds are studied, and systems of defining relations are derived for the braid groups of the plane and sphere. Chapter 2 focuses on the connections between the classical braid group and the classical knot problem. This is an area of research which has not progressed rapidly, yet there seem to be many interesting questions. The basic results are reviewed, and we then go on to prove an important theorem which was announced by Markov in 1935 but never proved in detail. This is followed by a discussion of a much newer result, Garside's solution to the conjugacy problem in the braid group. The last section of Chapter 2 explores some of the possible implications of the Garside and Markov theorems.

In Chapter 3 we discuss matrix representations of the free group and of subgroups of the automorphism group of the free group. These ideas come to a focus in the difficult open question of whether Burau's matrix representation of the braid group is faithful. In Chapter 4, we give an overview of recent results on the connections between braid groups and mapping class groups of surfaces. Finally, in Chapter 5, we discuss briefly the theory of "plats." The Appendix contains a list of problems. All are of a research nature, many of unknown difficulty.

It will be assumed that the reader is familiar with the basic ideas of elementary homotopy theory, such as the notions of covering spaces and fiber spaces, and of exact sequences of homotopy groups of pairs of spaces (a good reference is Hu's "Homotopy Theory," 1959); also, with elementary concepts in infinite group theory, such as the Schreier-Reidemeister rewriting process (see, for example, Magnus, Karass and Soliar, "Combinatorial Group Theory," 1966). With this qualification, we have attempted to make the manuscript self-contained.

On the matter of notation: Theorems are labeled consecutively within each chapter, e.g., Theorem 3.2 means the second theorem in Chapter 3; Corollary 3.2.2 means the second corollary to Theorem 3.2; Lemma 3.2.1 means the first lemma used in the proof of Theorem 3.2. Equations are numbered consecutively within each chapter, e.g., equation (3-33) means the thirty-third equation of Chapter 3. A double bar || is used to signify the end of a proof.

The suggestion that the lecture notes be the basis for a monograph originated with Ralph H. Fox. His lively interest and continuing encouragement, and his willingness to share completely the wealth of his knowledge and experience, did much to make this manuscript a reality.

I am deeply indebted to Charles F. Miller III, whose careful reading of the manuscript and many questions, criticisms and suggestions helped to make it both more readable and more accurate. The monograph was also reviewed by José María Montesinos; there is no adequate way to thank him for the time and effort and expertise which he brought to the task. The original lecture notes were taken by James Cannon, and I wish to thank him for his interest in the topics presented, and for the large amount of time and energy which he expended in the preparation of the notes. However, any errors which exist are certainly mine, because the manuscript has undergone extensive revisions from the original notes. I would also like to thank K. Murasugi, for numerous discussions about the possibility of applying braid theory to knots, which helped to clarify for me many of my own ideas. I am also grateful to all who attended the lectures, for their interest, questions and insights.

Finally, my thanks to J. H. Roberts, for communicating his unpublished proof of Theorem 4.4; to an unknown seminar speaker at Princeton University, circa 1954, for his notes on Theorem 2.3; and to the National Science Foundation of the United States for partial support.

JOAN S. BIRMAN

TABLE OF CONTENTS

Braids, Links, and

Mapping Class Groups

CHAPTER 1
BRAID GROUPS

The central theme in this manuscript is the concept of a braid group, and the many ways that the notion of a braid has been important in low dimensional topology. In particular, we will be interested in the largely unexplored possibility of applying braid theory to the study of knots and links, and also to the study of surface mappings.

Our object in this first chapter will be to develop the main structural and algebraic properties of braid groups on manifolds. (Our braid groups will be limited to groups of motions of points; we will not treat generalizations to motions of a sub-manifold of dimension > 0 in a manifold.) In this setting the "classical" braid group B_n of Artin appears as the "full braid group of the Euclidean plane E^2."

Section 1.1 is concerned with definitions. The problem of how to properly define a braid, in order to capture the essential significant properties of "weaving patterns" and so study them mathematically, is a very basic one, for if the definition is too narrow the range of application will be severely limited, while if it is too broad there will not be an interesting theory. It is a tribute to Artin's extraordinary insight as a Mathematician that the definition he proposed in 1925 [see Artin, 1925] for equivalence of geometric braids could ultimately be broadened and generalized in many different directions without destroying the essential features of the theory. For a discussion of several such generalizations, see Section 1.1; for generalizations to higher dimensions see [D. Dahm, 1962] and [D. Goldsmith, 1972]; for other generalizations, see [Brieskorn and Saito, 1972; Arnold, 1968b; Gorin and Lin, 1969].

Section 1.2 contains a development of the main properties of "configuration spaces," introduced by E. Fadell and L. Neuwirth in 1962. Configuration spaces will be our tool for finding defining relations in the braid groups of surfaces. We chose this method because we felt that it gave particular geometric insight into the algebraic structure of the classical braid group as a sequence of semi-direct products of free groups. This same structure is exhibited by other methods in [Magnus 1934; Markoff 1945; Chow 1948].

In Section 1.3 we review the chief properties of braid groups on manifolds other than E^2 and S^2. Theorem 1.5 shows that braid groups of manifolds M of dimension $n > 2$ are really not of much interest, since they are finite extensions (by the full symmetric group) of the nth cartesian product of $\pi_1 M$. Theorems 1.6 and 1.7 are concerned with the relationships between Artin's classical braid group on E^2 and the braid groups of other closed 2-manifolds.

In Section 1.4 we study the braid group of E^2. In Theorem 1.8 we find generators and defining relations for the full braid group B_n of E^2. Corollaries 1.8.1 and 1.8.2 relate to the algebraic structure of B_n as a sequence of semi-direct products of free groups, and lead to solutions to the "word problem" in B_n. In Corollary 1.8.3 we establish that B_n has a faithful representation as a subgroup of the automorphism group of a free group. This subgroup is characterized in Theorem 1.9, by giving necessary and sufficient conditions for an automorphism of a free group of rank n to be in B_n. Corollary 1.8.4 identifies the center of B_n. Finally, in Theorem 1.10 we establish another interpretation of B_n as the group of topologically-induced automorphisms of the fundamental group of an n-punctured disc, where admissible maps are required to keep the boundary of the disc fixed pointwise.

Section 1.5 discusses the braid group of the sphere, which will play an important role later in this book, in relation to the theory of surface mappings. In Section 1.6 we give a list of references for further results on braid groups of closed 2-manifolds.

1.1. *Definitions*

We begin not with the classical braid group, but with a somewhat more general concept of a braid as a motion of points in a manifold. Our definition will be shown to reduce to the classical case when the manifold is taken to be the Euclidean plane.

Let M be a manifold of dimension ≥ 2, let $\prod_{i=1}^{n} M$ denote the n-fold product space, and let $F_{0,n}M$ denote the subspace

$$F_{0,n}M = \left\{ (z_1,\cdots,z_n) \in \prod_{i=1}^{n} M / \ z_i \neq z_j \ \text{ if } \ i \neq j \right\}.$$

(The meaning of the subscript "0" in the symbol $F_{0,n}$ will become clear later.) The fundamental group $\pi_1 F_{0,n}M$ of the space $F_{0,n}M$ is the *pure* (or *unpermuted*) *braid group* with n strings of the manifold M.

Two points z and z' of $F_{0,n}M$ are said to be equivalent if the coordinates $(z_1,\cdots,z_n)$ of z differ from the coordinates $(z'_1,\cdots,z'_n)$ of z' by a permutation. Let $B_{0,n}M$ denote the identification space of $F_{0,n}M$ under this equivalence relation. The fundamental group $\pi_1 B_{0,n}M$ of the space $B_{0,n}M$ is called the *full braid group* of M, or more simply, the *braid group* of M. Note that the natural projection $\rho : F_{0,n}M \to B_{0,n}M$ is a regular covering projection.

The classical braid group of Artin [cf. Artin 1925 and 1947a] is the braid group $\pi_1 B_{0,n}E^2$, where E^2 denotes the Euclidean plane. Artin's geometric definition of $\pi_1 B_{0,n}E^2$ can be recovered from the definition above as follows:

Choose a base point $z^0 = (z_1^0,\cdots,z_n^0) \in F_{0,n}E^2$ for $\pi_1 F_{0,n}E^2$ and a point $\tilde{z}^0 \in B_{0,n}E^2$ such that $\rho(z^0) = \tilde{z}^0$. Any element in $\pi_1 B_{0,n}E^2 = \pi_1(B_{0,n}E^2, \tilde{z}^0)$ is represented by a loop

$$\ell : I,\{0,1\} \to B_{0,n}E^2, \tilde{z}^0$$

which lifts uniquely to a path

$$\ell : I,\{0\} \to F_{0,n}E^2, z^0.$$

If $f(t) = (f_1(t), \cdots, f_n(t))$, $t \in I$, then each of the coordinate functions f_i defines (via its graph) an arc $\mathcal{A}_i = (f_i(t), t)$ in $E^2 \times I$. Since $f(t) \in F_{0,n}E^2$ the arcs $\mathcal{A}_1, \cdots, \mathcal{A}_n$ are disjoint. Their union $\mathcal{A} = \mathcal{A}_1 \cup \cdots \cup \mathcal{A}_n$ is called a *geometric braid* (see Figure 1). The arc $\mathcal{A}_i$ is called the ith *braid string.*

A geometric braid is a representative of a path class in the fundamental group $\pi_1 B_{0,n}E^2$. Thus if $\mathcal{A}$ and $\mathcal{A}'$ are geometric braids, then $\mathcal{A} \sim \mathcal{A}'$ (that is, they represent the same element of $\pi_1 B_{0,n}$) if the paths f and f' which define these braids are homotopic relative to the base point $(z_1^0, \cdots, z_n^0)$ in the space $F_{0,n}E^2$. Thus we require the existence of a continuous mapping $\mathcal{F} : I \times I \to F_{0,n}E^2$ with

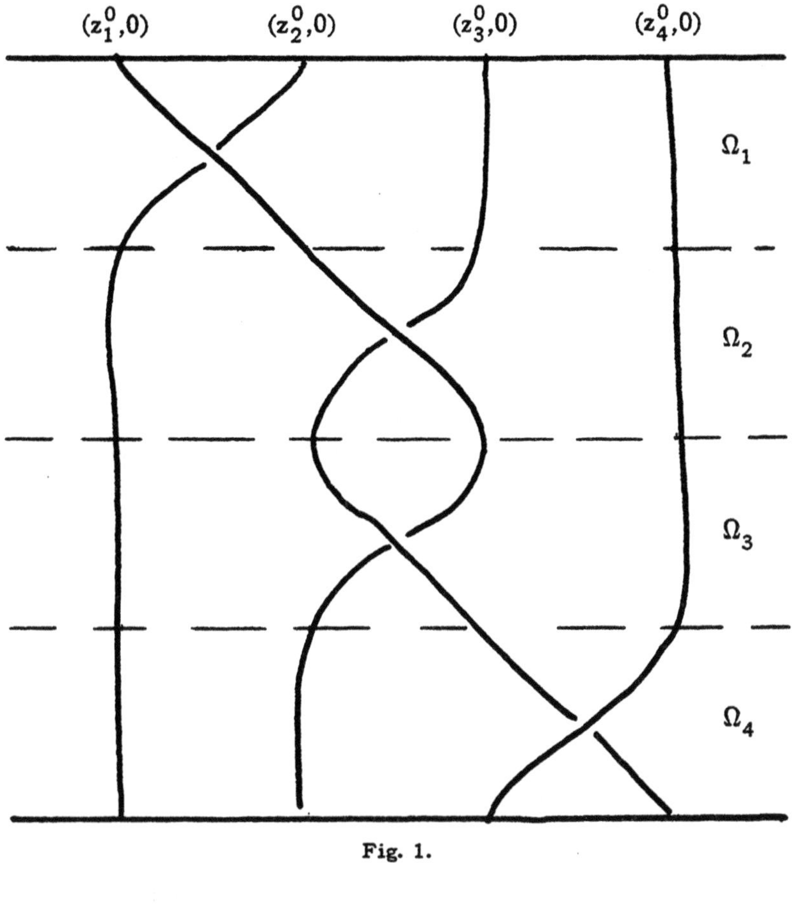

Fig. 1.

$$\mathcal{F}(t,0) = (\mathcal{F}_1(t,0),\cdots,\mathcal{F}_n(t,0)) = (\ell_1(t),\cdots,\ell_n(t))$$

$$\mathcal{F}(t,1) = (\mathcal{F}_1(t,1),\cdots,\mathcal{F}_n(t,1)) = (\ell'_1(t),\cdots,\ell'_n(t))$$

$$\mathcal{F}(0,s) = (\mathcal{F}_1(0,s),\cdots,\mathcal{F}_n(0,s)) = (z_1^0,\cdots,z_n^0)$$

$$\mathcal{F}(1,s) = (\mathcal{F}_1(1,s),\cdots,\mathcal{F}_n(1,s)) = (z_{\mu_1}^0,\cdots,z_{\mu_n}^0)$$

where $(\mu_1,\cdots,\mu_n)$ is a permutation of the array $(1,\cdots,n)$. The homotopy $\mathcal{F}$ defines a continuous sequence of geometric braids $\mathfrak{A}(s) = \mathfrak{A}_1(s)\cup\cdots\cup \mathfrak{A}_n(s)$, $s \in I$, where $\mathfrak{A}_i(s) = (\mathcal{F}_i(t,s),t)$, such that $\mathfrak{A}(0) = \mathfrak{A}$ and $\mathfrak{A}(1)=\mathfrak{A}'$. The reader is referred to Figures 2(b) and 2(c) for pictures of geometric braids which are equivalent to the "trivial" braid.

One may also define various stronger and weaker forms of equivalence between geometric braids, and we mention several of these briefly:

i). Let $\mathfrak{A}$ and $\mathfrak{A}'$ be geometric braids. Note that $\mathfrak{A}$ and $\mathfrak{A}'$ are subsets of $E^2 \times I$. Then, we write $\mathfrak{A} \approx \mathfrak{A}'$ if there is an isotopic deformation $\mathcal{G}_s$ of $E^2 \times I$ which is the identity on $E^2 \times \{0\}$ and on $E^2 \times \{1\}$ for each $s \in [0,1]$ and which has the property:

*For each $s \in [0,1]$ the image set $\mathfrak{A}(s)$ of $\mathfrak{A}$ under $\mathcal{G}_s$ is a geometric braid, that is, $\mathfrak{A}(s)$ meets each plane $E^2 \times \{t_0\}$, $t_0 \in I$, in precisely n points, and moreover $\mathfrak{A}(0) = \mathfrak{A}$, $\mathfrak{A}(1) = \mathfrak{A}'$.

It was proved by Artin [see 1947a] that $\mathfrak{A} \approx \mathfrak{A}'$ if and only if $\mathfrak{A} \sim \mathfrak{A}'$. Thus a braid homotopy may always be "extended" to $E^2 \times I$, in the sense defined above.

ii). If we think of our braid strings $\mathfrak{A}_1,\cdots,\mathfrak{A}_n$ as being made of elastic, one might imagine a more general type of equivalence in which the strings could be stretched or deformed in the region $E^2 \times I$ *without* requiring that $\mathfrak{A}(s)$ meet each plane $E^2 \times \{t_0\}$, $t_0 \in I$, in precisely n points. In this situation, it might happen, for example, that some intermediate set $\mathfrak{A}_1(s_0)\cup\cdots\cup\mathfrak{A}_n(s_0)$ is as illustrated in Figure 3. (This intermediate set is not a geometric braid.) More precisely, under this

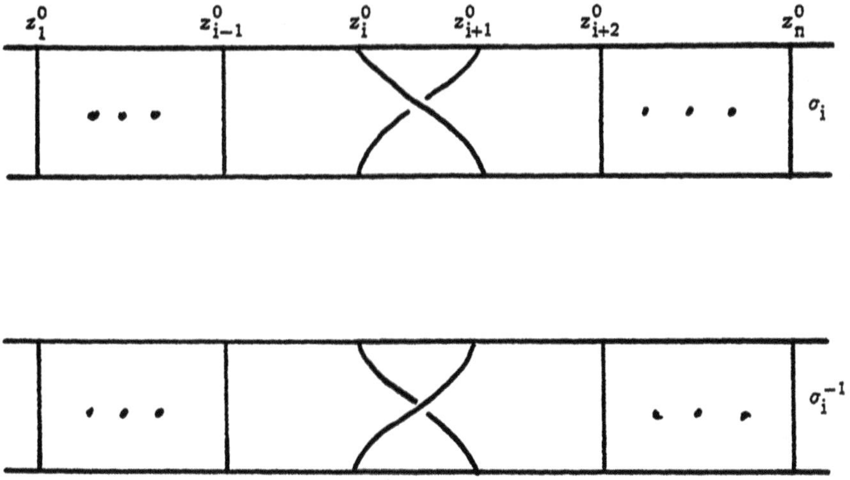

(a) Geometric braids representing σ_i and σ_i^{-1}

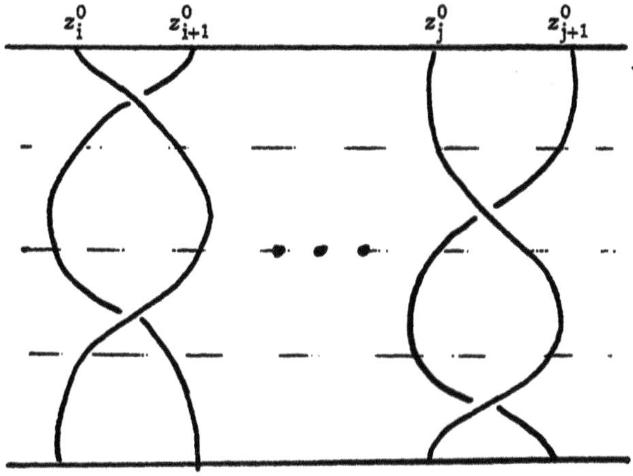

(b) $\sigma_i \sigma_j \sigma_i^{-1} \sigma_j^{-1} \approx$ identity if $|i-j| > 2$

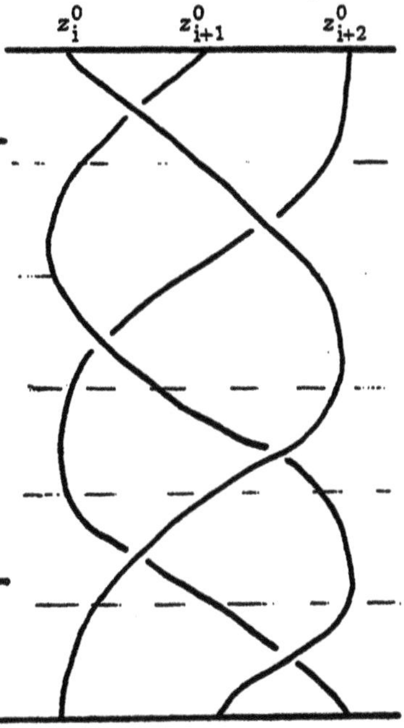

(c) $\sigma_i \sigma_{i+1} \sigma_i \sigma_{i+1}^{-1} \sigma_i^{-1} \sigma_{i+1}^{-1} \approx$ identity

Fig. 2.

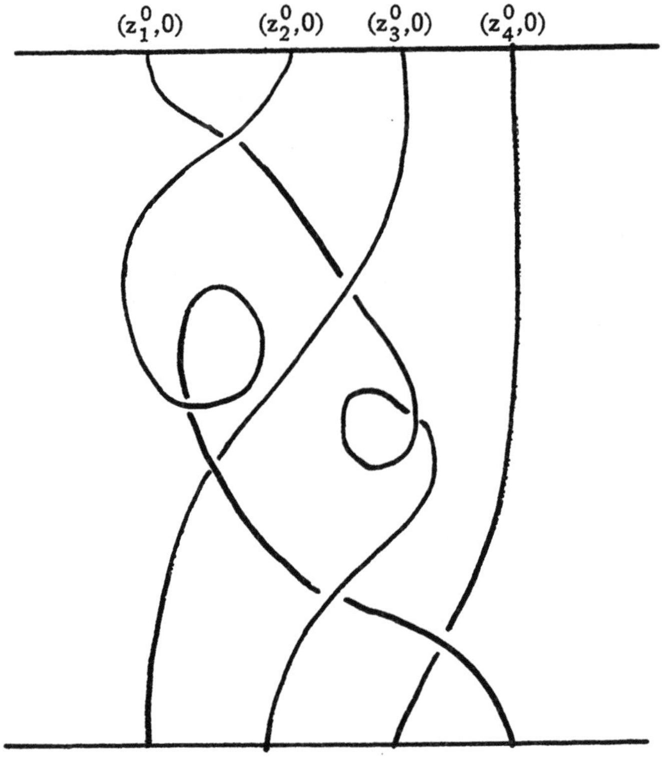

Fig. 3

more general notion, $\mathfrak{A} \approx \mathfrak{A}'$ if there is an isotopy $\mathcal{G}_s$ which is exactly like that defined in i) above except that $\mathcal{G}_s$ need not satisfy the property $*$. Again, Artin established [1947a] that $\mathfrak{A} \approx \mathfrak{A}'$ if and only if $\mathfrak{A} \sim \mathfrak{A}'$. (This is the first hint of a relationship between the concepts of equivalence of braids and equivalence of links, a relationship which will be studied in detail in Chapter 2.)

iii). D. Goldsmith [1974] has defined a concept of "homotopy" of braids by defining two geometric braids to be homotopic if one can be deformed to the other by simultaneous homotopies of the individual paths $(f_i(t), t)$ in $E^2 \times I$, fixing the end points, and subject to the restriction that a string may intersect *itself*, but not any other string. Note that if $\mathfrak{A} \sim \mathfrak{A}'$ then $\mathfrak{A}$ and $\mathfrak{A}'$ are also equivalent under Goldsmith's rule, but the converse need not be true. In fact, Goldsmith has exhibited non-trivial elements of the group $\pi_1 B_{0,3}E^2$ which are homotopic to the identity element of $\pi_1 B_{0,3}E^2$. She goes on to define a "homotopy braid group," $\hat{B}_n$, and finds a group presentation for $\hat{B}_n$ which exhibits $\hat{B}_n$ as a quotient group of the group $\pi_1 B_{0,n}E^2$. We note that Goldsmith's results were suggested by J. Milnor's work on homotopy of links and isotopy of links [see J. Milnor, 1954].

iv). The concept of a braid group has been generalized by D. Dahm [1962] and by D. Goldsmith [1972] to a group of motions of a submanifold in a manifold. We now give Goldsmith's definition of that group. Let N be a subspace contained in the interior of a manifold M. Denote by $\mathcal{H}(M)$ the group of autohomeomorphisms of M with the compact open topology, where if M has boundary ∂M, all homeomorphisms are required to fix ∂M pointwise. Denote the identity map of M by $1_M : M \to M$. A *motion of* N *in* M is a path f_t in $\mathcal{H}(M)$ beginning at $f_0 = 1_M$ and ending at f_1, where $f_1(N) = N$. The motion is said to be a *stationary motion of* N *in* M if $f_t(N) = N$ for all $t \in [0,1]$. To compose two motions, translate the second by multiplication in the group $\mathcal{H}(M)$ so that its initial point coincides with the endpoint of the first, and multiply as in the groupoid of paths. Define the inverse f^{-1} of a motion f to be the inverse of the path f in $\mathcal{H}(M)$, translated so that its initial point is 1_M.

Finally, let motions f and g be equivalent if the path $f^{-1}g$ is homotopic modulo its endpoint to a stationary motion. The *group of motions of* N *in* M is the set of equivalence classes of motions of N in M, with multiplication induced by composition of motions. From this point of view, the group of motions of an interior point in a manifold M is the group $\pi_1 M$, and the group of motions of n distinct points is the pure braid group of M (cf. Chapter 4 of this text, also Theorem 1.10).

Dahm [1962] studies the group of motions of n disjoint circles in S^3, and Goldsmith [1972, studies the group of motions of torus links in S^3.

1.2. *Configuration spaces*

PROPOSITION 1.1. *The natural projection map* $p: F_{0,n}M \to B_{0,n}M$ *is a regular covering space projection. The group of covering transformations is the full symmetric group* Σ_n *on* n *letters. Therefore there is a canonical isomorphism*

$$\text{(1-3)} \qquad \pi_1 B_{0,n}M/\pi_1 F_{0,n}M \approx \Sigma_n .$$

The geometric interpretation of the product of two braids is immediate; suffice it to say that, in Figure 1, the configuration of arcs between any two consecutive dotted horizontal levels can be considered to be geometric braids $(\Omega_1, \Omega_2, \Omega_3, \Omega_4)$ of which the entire braid $\Omega = \Omega_1\Omega_2\Omega_3\Omega_4$ is the product.

Geometric intuition suggests that an arbitrary braid is equivalent to a braid that is a product of simple braids of the types illustrated in Figure 2. The equivalence classes of these elementary braids will be denoted by the symbols σ_i and σ_i^{-1}. In the example of Figure 1,

$$\Omega = \sigma_1 \sigma_2^2 \sigma_3^{-1} .$$

Geometric intuition thus suggests that $\sigma_1, \cdots, \sigma_{n-1}$ generate the group $\pi_1 B_{0,n}E^2$, a fact which will be proved later.

The following relations in $\pi_1 B_{0,n}E^2$ are obvious from Figure 2:

$$\text{(1-1)} \qquad \sigma_i\sigma_j = \sigma_j\sigma_i \quad \text{if } |i-j| \geq 2,\ 1 \leq i, j \leq n-1$$

$$\text{(1-2)} \qquad \sigma_i\sigma_{i+1}\sigma_i = \sigma_{i+1}\sigma_i\sigma_{i+1} \qquad 1 \leq i \leq n-2 .$$

It will be proved below that (1-1) and (1-2) comprise a set of defining relations in $\pi_1 B_{0,n}E^2$. Our proof, which allows us at the same time to compute defining relations for the braid groups of arbitrary 2-manifolds, will make use of the concept of the "configuration space" of a manifold. Other proofs (for the special case $M = E^2$) can be found in [Artin 1925; 1947a; Magnus 1934; Bohnenblust 1947; Fox and Neuwirth 1962].

Proof. Clear. ||

Since the map p is known explicitly, it follows from Proposition 1.1 that it is not difficult to analyze $\pi_1 B_{0,n}M$ once $\pi_1 F_{0,n}M$ is known. Therefore, the remainder of this section will be devoted to the group $\pi_1 F_{0,n}M$.

Let $Q_m = \{q_1, \cdots, q_m\}$ be a set of fixed distinguished points of M. Following Fadell and Neuwirth [1962] and Fadell and Van Buskirk [1962] we define the *configuration space* $F_{m,n}M$ of M to be the space $F_{0,n}(M - Q_m)$. Note that the topological type of $F_{m,n}M$ does not depend on the choice of the particular points Q_m, since one may always find an isotopy of M which deforms any one such point set Q_m into any other Q'_m. Note that $F_{m,1}M = M - Q_m$. (One may, similarly, define spaces $B_{m,n}M = B_{0,n}(M - Q_m)$, however we will only be interested in $B_{0,n}M$.)

We are interested in the relationship between the configuration spaces $F_{n,m}M$ and $F_{0,n}M$. The key observation is the following theorem:

THEOREM 1.2 [Fadell and Neuwirth, 1962]. *Let* $\pi : F_{m,n}M \to F_{m,r}M$ *be defined by*

(1-4) $$\pi(z_1, \cdots, z_n) = (z_1, \cdots, z_r), \quad 1 \le r < n .$$

Then π *exhibits* $F_{m,n}M$ *as a locally trivial fibre space over the base space* $F_{m,r}M$, *with fibre* $F_{m+r,n-r}M$.

Proof. First consider, for some base point $(z_1^0, \cdots, z_r^0)$ in $F_{m,r}M$, the fibre $\pi^{-1}(z_1^0, \cdots, z_r^0)$:

$$\pi^{-1}(z_1^0, \cdots, z_r^0) = \{(z_1^0, \cdots, z_r^0, y_{r+1}, \cdots, y_n), \text{ where } z_1^0, \cdots, z_r^0, y_{r+1}, \cdots, y_n \text{ are distinct and in } M - Q_m\}.$$

If we select Q_{m+r} equal to $Q_m \cup \{z_1^0, \cdots, z_r^0\}$, then

$$F_{m+r,n-r}M = \{(y_{r+1}, \cdots, y_n), \text{ where } y_{r+1}, \cdots, y_n \text{ are distinct and in } M - Q_{m+r}\},$$

and there is an obvious homeomorphism

$$h : F_{m+r,n-r}M \to \pi^{-1}(z_1^0,\cdots,z_r^0)$$

defined by

$$h(y_{r+1},\cdots,y_n) = (z_1^0,\cdots,z_r^0, y_{r+1},\cdots,y_n)\,.$$

The proof of the local triviality of π will be carried out, for notational and descriptive convenience, only in the case of $r = 1$. The other cases will be left to the reader as exercises. Fix for consideration, therefore, a point $x_0 \in M - Q_m = F_{m,1}M = F_{m,r}M$. Add another point q_{m+1} to the set Q_m to form Q_{m+1} and pick a homeomorphism $\alpha : M \to M$, fixed on Q_m, such that $\alpha(q_{m+1}) = x_0$. Let U denote a neighborhood of x_0 in $M - Q_m$ which is homeomorphic to an open ball, and let $\bar{U}$ denote the closure of U. Define a map $\theta : U \times \bar{U} \to \bar{U}$ with the following properties. Setting $\theta_z(y) = \theta(z,y)$ we require:

(i) $\theta_z : \bar{U} \to \bar{U}$ is a homeomorphism which fixes $\partial\bar{U}$.

(ii) $\theta_z(z) = x_0$.

By (i), θ can be extended to $\theta : U \times M \to M$ by defining $\theta(z,y) = y$ for $y \notin U$. The required local product representation

$$U \times F_{m+1,n-1}M \underset{\phi^{-1}}{\overset{\phi}{\rightleftarrows}} \pi^{-1}(U)$$

is given by

$$\phi(z,z_2,\cdots,z_n) = (z,\theta_z^{-1}\alpha(z_2),\cdots,\theta_z^{-1}\alpha(z_n))\,.$$

$$\phi^{-1}(z,z_2,\cdots,z_n) = (z,\alpha^{-1}\theta_z(z_2),\cdots,\alpha^{-1}\theta_z(z_n))\,. \;\|$$

Two important consequences of Theorem 1.2 now follow:

PROPOSITION 1.3. *If* $\pi_2(M - Q_m) = \pi_3(M - Q_m) = 0$ *for each* $m \geq 0$, *then* $\pi_2 F_{0,n}M = 0$.

Proof. The exact homotopy sequence of the fibration $\pi : F_{m,n}M \to F_{m,1}M = M - Q_m$ of Theorem 1.2 gives an exact sequence

$$\cdots \to \pi_3(M-Q_m) \to \pi_2 F_{m+1,n-1}M \to \pi_2 F_{m,n}M \to \pi_2(M-Q_m) \to \cdots .$$

Since $\pi_2(M-Q_m) = \pi_3(M-Q_m) = 0$, it follows that $\pi_2 F_{m+1,n-1}M$ and $\pi_2 F_{m,n}M$ are isomorphic. An inductive argument shows that

$$\pi_2 F_{0,n}M \approx \pi_2 F_{n-1,1}M = \pi_2(M-Q_{n-1}) = 0 . \tag{1-5}$$

This completes the proof. ||

Let π be the projection map from $F_{0,n}M$ to $F_{0,n-1}M$ defined by (1-4). Let $(z_1^0,\cdots,z_n^0)$ be base point for $\pi_1 F_{0,n}M$. Let $F_{n-1,1}M = M - Q_{n-1} = M - \{z_1^0,\cdots,z_{n-1}^0\}$. Let j be the inclusion map from $F_{n-1,1}M$ to $F_{0,n}M$, defined by

$$j(z_n) = (z_1^0,\cdots,z_{n-1}^0, z_n) \qquad z_n \in M - \{z_1^0,\cdots,z_{n-1}^0\} . \tag{1-6}$$

THEOREM 1.4. *If* $\pi_2(M-Q_m) = \pi_3(M-Q_m) = \pi_0(M-Q_m) = 1$ *for every* $m \geq 0$, *then the following sequence of groups and homomorphism is exact:*

$$1 \longrightarrow \pi_1(F_{n-1,1}M, z^0) \xrightarrow{j_*} \pi_1(F_{0,n}M, (z_1^0,\cdots,z_n^0)) \tag{1-7}$$

$$\xrightarrow{\pi_*} \pi_1(F_{0,n-1}M, (z_1^0,\cdots,z_{n-1}^0)) \longrightarrow 1$$

where π_* *and* j_* *are the homomorphism induced by the mappings* π *and* j.

Proof. The sequence (1-7) is part of the exact homotopy sequence of the fibration of Theorem 1.2. The identity terms reflect the equalities $\pi_2 F_{0,n-1} = 1$, established in Proposition 1.3, and $\pi_0 F_{n-1,1}M = \pi_0(M-Q_{n-1}) = 1$. ||

The exact sequence (1-7) will be used later, in conjunction with Proposition 1.1, to determine group presentations for $\pi_1 B_{0,n}E^2$ and $\pi_1 B_{0,n}S^2$.

1.3. *Braid groups of manifolds*

An indication that the most interesting braid groups are those on 2-dimensional manifolds is given by the following theorem:

THEOREM 1.5 [Birman, 1969a, pp. 42-44]. *Let* M *be a closed, smooth manifold of dimension* n. *Then for each integer* k *the inclusion map* $i_k: F_{0,n}M \to \prod_n M$ *induces a homomorphism*

$$(i_k)_* : \pi_k F_{0,n}M \to \prod_n \pi_k M \tag{1-8}$$

which is surjective if $\dim M > k$ *and also injective if* $\dim M > 1 + k$.

Since our interest here is primarily in surfaces, we will not prove Theorem 1.5. The reader is referred to [Birman, 1969a], or to [Dahm, 1962], for proofs.

Among the braid groups on 2-dimensional manifolds, Artin's classical braid group $\pi_1 B_{0,n}E^2$ and Artin's pure braid group $\pi_1 F_{0,n}E^2$ hold central positions. This assertion is justified by the remarks which follow, and by Theorems 1.6 and 1.7 below, which are based on material in [Birman, 1969a] and in [Goldberg, 1973].

Let $(z_1^0, \cdots, z_n^0)$ be the base point for the group $\pi_1 F_{m,n}E^2$, as before, and regard E^2 as an open disc in M which contains the n points $z_1^0, \cdots, z_n^0$ and also the distinguished set Q_m. Let $P_n = (p_1, \cdots, p_n)$ be a fixed n-point set for each n. Then we may regard $F_{m,n}M$ as the set of embeddings of P_n in $(M - Q_m)$. From this point of view, the space

$F_{m,n}E^2$ may be identified with a subset of $F_{m,n}M$ by composing any map from P_n to $F_{m,n}E^2$ with the inclusion map $E^2 \subseteq M$. Let $\epsilon_{m,n}: F_{m,n}E^2 \to F_{m,n}M$ be the resulting identification. Then the induced map $\epsilon^*_{m,n}: \pi_1 F_{m,n}E^2 \to \pi_1 F_{m,n}M$ takes any n-string braid on $(E^2 - Q_m)$ and considers it as a braid in $(M - Q_m)$.

THEOREM 1.6. *If* M *is any compact surface except* S^2 *or* P^2, *then* $\ker \epsilon^*_{0,n} = \{1\}$.

Proof [Birman, 1969a; see also Goldberg, 1973 for a different proof]. The homomorphism $\epsilon^*_{m,n}$ together with the exact sequences of Theorem 1.4 yield a commutative diagram:

$$\begin{array}{ccccccccc}
1 & \longrightarrow & \pi_1(E^2-Q_{n-1}) & \longrightarrow & \pi_1 F_{0,n}E^2 & \longrightarrow & \pi_1 F_{0,n-1}E^2 & \longrightarrow & 1 \\
 & & \downarrow \epsilon^*_{n-1,1} & & \downarrow \epsilon^*_{0,n} & & \downarrow \epsilon^*_{0,n-1} & & \\
1 & \longrightarrow & \pi_1(M-Q_{n-1}) & \longrightarrow & \pi_1 F_{0,n}M & \longrightarrow & \pi_1 F_{0,n-1}M & \longrightarrow & 1
\end{array}$$

Note that $\epsilon^*_{n-1,1}$ is injective for each n. This fact and the diagram may be used inductively to establish that $\epsilon^*_{0,n}$ is injective as well. Clearly $\epsilon^*_{0,1}$ is injective since $\pi_1 F_{0,1}E^2 = \pi_1 E^2 = 1$. This begins the induction. Suppose inductively that $\epsilon^*_{0,n-1}$ is injective. Then the strong 5-lemma implies that $\epsilon^*_{0,n}$ is injective. This completes the induction and the proof of the proposition. ||

Focusing on the braid group $\pi_1 F_{0,n}M$ of a compact 2-manifold M, one readily discerns two distinct types of phenomena which are exhibited by representatives of elements of the group $\pi_1 F_{0,n}M$.

i) There is "classical braiding," which may be thought of as taking place in the open disc $E^2 \subseteq M$.

ii) There is wandering of the individual strands about on the surface M. The next theorem says that, in effect, for a closed surface $M \neq S^2$ or P^2, nothing else happens.

THEOREM 1.7 [Goldberg, 1973]. *Let* M *be a closed surface different from* S^2 *or* P^2. *Let* $i: F_{0,n}M \to \prod_n M$ *be the inclusion map. Then in the following sequence of (not necessarily abelian) groups*

$$1 \longrightarrow \pi_1 F_{0,n}E^2 \xrightarrow{e_{0,n}^*} \pi_1 F_{0,n}M \xrightarrow{i_*} \prod_{i=1}^{n} \pi_1 M \longrightarrow 1$$

the kernel of each homomorphism is equal to the normal closure of the image of the previous homomorphism in the sequence.

Proof. The reader is referred to Goldberg's paper for the proof of Theorem 1.7. ||

1.4. *The braid group of the plane*

In view of Theorem 1.5, Proposition 1.6 and Theorem 1.7 it is apparent that the groups $\pi_1 B_{0,n}E^2$ and $\pi_1 F_{0,n}E^2$ merit special attention. In this section we will consider only the case $M = E^2$. Accordingly, abbreviations $B_{0,n}$ and $F_{0,n}$ will be adopted for the spaces $B_{0,n}E^2$ and $F_{0,n}E^2$.

In this section the short exact sequence of Theorem 1.4 will be used inductively to show that $\pi_1 F_{0,n}$ is constructed in nice ways from the building blocks $\pi_1(E^2-Q_i)$, $1 \leq i \leq n-1$ (Corollary 1.8.1), finding simultaneously generators and defining relations for the groups $\pi_1 B_{0,n}$ and $\pi_1 F_{0,n}$ (Theorem 1.8 and Lemma 1.8.1). From the structure of the group $\pi_1 F_{0,n}$ (uncovered in Corollary 1.8.1) a unique normal form is developed, in Corollary 1.8.2, for elements in $\pi_1 B_{0,n}$. This leads to a solution to the word problem in $\pi_1 B_{0,n}$. In Corollary 1.8.3 it is proved that $\pi_1 B_{0,n}$ has a faithful representation as a group of automorphisms of a free group of rank n. In Theorem 1.9 the particular subgroup of the automorphism group of a free group which is so obtained is characterized algebraically. Theorem 1.10 gives a new geometric meaning to the group $\pi_1 B_{0,n}$.

THEOREM 1.8 [Artin, 1925]. *The group* $\pi_1 B_{0,n}$ *admits a presentation with generators* $\sigma_1, \cdots, \sigma_{n-1}$ *and defining relations*

$$\text{(1-1)} \qquad \sigma_i \sigma_j = \sigma_j \sigma_i \quad \textit{if} \quad |i-j| \geq 2,\ 1 \leq i, j \leq n-1$$

$$\text{(1-2)} \qquad \sigma_i \sigma_{i+1} \sigma_i = \sigma_{i+1} \sigma_i \sigma_{i+1} \qquad 1 \leq i \leq n-2\ .$$

Proof [The proof given here is due to Fadell and Van Buskirk, 1962]. Let B_n be the abstract group with the presentation of Theorem 1.8. Until we have established the isomorphism between B_n and $\pi_1 B_{0,n}$, we will use the symbols $\tilde{\sigma}_1, \cdots, \tilde{\sigma}_{n-1}$ for elements of $\pi_1 B_{0,n}$ with $\iota : B_n \to \pi_1 B_{0,n}$ defined by $\iota(\sigma_i) = \tilde{\sigma}_i$, $1 \leq i \leq n-1$. The elements $\tilde{\sigma}_i$ were already defined by pictures in Figure 2. (Anticipating the result of Theorem 1.8, we used the symbols σ_i and σ_i^{-1} in Figure 2.) We now give an equivalent definition which is more precise. Recall the covering projection $p : F_{0,n} \to B_{0,n}$. Choose the point $p((1,0), \cdots, (n,0)) = \tilde{z}^0$ as base point for the group $\pi_1 B_{0,n}$. Lift loops based at $p((1,0), \cdots, (n,0))$ in $B_{0,n}$ to paths in $F_{0,n}$ with initial point $((1,0), \cdots, (n,0)) = \tilde{\tilde{z}}^0$. Then the generator $\tilde{\sigma}_i \in \pi_1 B_{0,n}$ is represented by the path $\ell(t)$ in $F_{0,n}$ given by

$$\text{(1-9)} \qquad \ell(t) = ((1,0), \cdots, (i-1,0), \ell_i(t), \ell_{i+1}(t), (i+2,0), \cdots, (n,0))\ ,$$

where $\ell_i(t) = (i+t, -\sqrt{t-t^2})$ and $\ell_{i+1}(t) = (i+1-t, \sqrt{t-t^2})$. That is, $\ell(t)$ is constant on all but the ith and $i+1^{st}$ strings and interchanges those two in a nice way.

The proof of Theorem 1.8 will be by induction on n, and will exploit the relationship already developed in Proposition 1.1 between $\pi_1 B_{0,n}$

and $\pi_1 F_{0,n}$. Let

$$\tilde{\nu} : \pi_1(B_{0,n}, \tilde{z}^0) \to \Sigma_n$$

be defined as follows: Let $\tilde{a} \in \pi_1 B_{0,n}$ be represented by a loop

$$\tilde{g} : (I, \{0,1\}) \to (B_{0,n}, \tilde{z}^0)$$

and let $g = (g_1, \cdots, g_n) : (I, \{0\}) \to (F_{0,n}, z^0)$ be the unique lift of $\tilde{g}$. Define

$$\tilde{\nu}(a) = \begin{pmatrix} g_1(0), \cdots, g_n(0) \\ g_1(1), \cdots, g_n(1) \end{pmatrix} \in \Sigma_n .$$

The kernel of the homomorphism $\tilde{\nu}$ is the pure braid group, $\pi_1 F_{0,n}$. Corresponding to the homomorphism $\tilde{\nu}$ is the homomorphism

$$\nu : B_n \to \Sigma_n$$

from the abstract group B_n to the symmetric group Σ_n on n letters defined by:

(1-10) $\qquad \nu(\sigma_i) = (i, i+1) \qquad 1 \leq i \leq n-1 .$

Let $P_n = \ker \nu$.

LEMMA 1.8.1. *The homomorphism* $\iota : B_n \to \pi_1 B_{0,n}$ *is an isomorphism onto if* $\iota | P_n$ *is an isomorphism onto* $\pi_1 F_{0,n}$.

Proof of Lemma 1.8.1. The homomorphism ν is clearly surjective, since the transpositions $\{\nu(\sigma_i)\ 1 \leq i \leq n-1\}$ generate Σ_n. Hence we have a commutative diagram

$$\begin{array}{ccccccccc} 1 & \longrightarrow & P_n & \longrightarrow & B_n & \xrightarrow{\nu} & \Sigma_n & \longrightarrow & 1 \\ & & \downarrow{\scriptstyle \iota_n = i|P_n} & & \downarrow{\scriptstyle \iota} & & \downarrow{\scriptstyle 1} & & \\ 1 & \longrightarrow & \pi_1 F_{0,n} & \longrightarrow & \pi_1 B_{0,n} & \xrightarrow{\tilde{\nu}} & \Sigma_n & \longrightarrow & 1 \end{array}$$

with exact rows. Applying the "Five Lemma,"[1] we obtain the desired result. This completes the proof of Lemma 1.8.1. ||

In order to demonstrate that $i|P_n$ is an isomorphism onto, we next find a presentation for P_n.

LEMMA 1.8.2. *The group* P_n *admits a presentation with generators*

$$(1\text{-}11) \quad A_{ij} = \sigma_{j-1}\sigma_{j-2}\cdots\sigma_{i+1}\sigma_i^2\sigma_{i+1}^{-1}\cdots\sigma_{j-2}^{-1}\sigma_{j-1}^{-1} \quad (1 \le i < j \le n)$$

and defining relations[2]

$$\begin{aligned}(1\text{-}12) \quad A_{rs}A_{ij}A_{rs}^{-1} &= A_{ij} \quad \textit{if} \quad r < s < i < j \ \text{ or } \ i < r < s < j .\\ &= A_{rj}A_{ij}A_{rj}^{-1} \quad \textit{if} \quad s = i .\\ &= A_{ij}A_{sj}A_{ij}A_{ij}A_{sj}^{-1} \quad \textit{if} \quad r = i < j < s .\\ &= A_{rj}A_{sj}A_{rj}^{-1}A_{sj}^{-1}A_{ij}A_{rj}A_{sj}A_{rj}^{-1}A_{sj}^{-1\,-1}\\ &\quad \textit{if} \quad r < i < s < j .\end{aligned}$$

(The reader is referred to Figure 4 for a picture of the geometric braid A_{ij}.)

Proof of Lemma 1.8.2. The group P_n is of index $n!$ in B_n. We may choose as coset representatives for P_n in B_n any set of $n!$ words in the generators of B_n whose images under ν range over all of Σ_n. In particular, a set of right coset representatives are the collection of all products of the form $\{\prod_{j=2}^{n} M_{j,k_j};\ j \ge k_j \ge 1\}$ where $M_{j,i} = \sigma_{j-1}\sigma_{j-2}\cdots\sigma_i$

---

1 See, for example, Eilenberg and Steenrod, "Foundations of Algebraic Topology," Princeton Univ. Press (1952), p. 16.

2 Note that our relations are slightly different from, but equivalent to, the system of relations found by Artin, 1947a, p. 120. Our method is different from Artin's. The presentation of Artin, 1947a, is reproduced in Magnus, Karass and Solitar, 1966, p. 174. We remark that the relations given by Artin on p. 120 of his 1947a paper are correct if $\varepsilon = +1$, but incorrect if $\varepsilon = -1$. The relations given in Magnus, Karass and Solitar are based on Artin's relations for the case $\varepsilon = +1$, and are correct.

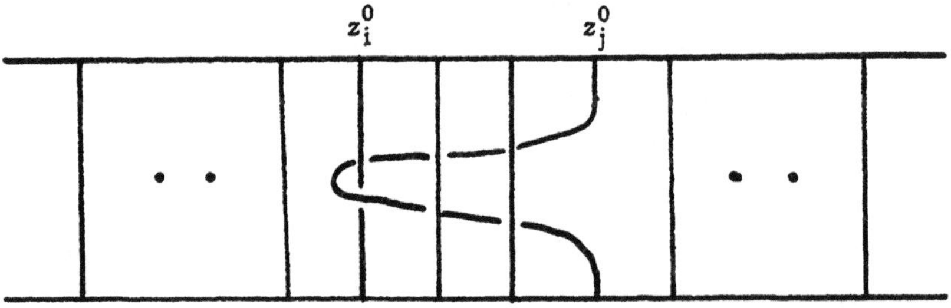

(a) The geometric braid A_{ij}

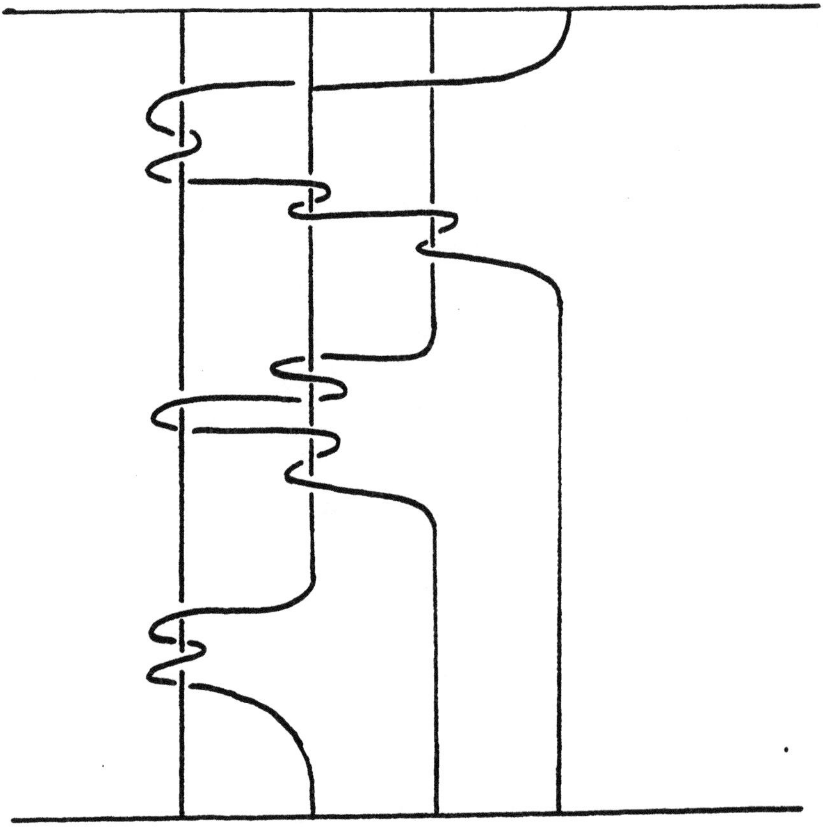

(b) Combed 4-braid

Fig. 4.

if $j \neq i$, or 1 if $j = i$. Thus, for example, coset representatives for P_3 in B_3 are the set $M_{22}M_{33} = 1$, $M_{22}M_{32} = \sigma_2$, $M_{22}M_{31} = \sigma_2\sigma_1$, $M_{21}M_{33} = \sigma_1$, $M_{21}M_{32} = \sigma_1\sigma_2$, and $M_{21}M_{31} = \sigma_1\sigma_2\sigma_1$. This is a Schreier set; that is, any initial segment of a coset representative is again a coset representative. Hence we may apply the Schreier-Reidemeister method [see, for example, Magnus, Karass and Solitar, 1966, Section 2.3] to obtain a group presentation for P_n.

For the reader who wishes to verify our presentation, we note that an alternative method of proving Lemma 1.8.2, which is conceptually more complex than that described above, but mechanically easier to handle, is given by [Chow, 1948]. Chow's approach is to treat the calculation in stages, by first obtaining a presentation for the subgroup D_n of all elements in B_n which have permutations which leave the letter n invariant. A Schreier set of coset representatives for D_n in B_n are the set $\{M_{n,k_n};\ n \geq k_n \geq 1\}$. The Schreier-Reidemeister method applied to D_n exhibits D_n as the semi-direct product of the normal subgroup U_n generated by $A_{1n}, A_{2n}, \cdots, A_{n-1,n}$ and the subgroup of $D_n \subseteq B_n$ generated by $\sigma_1, \cdots, \sigma_{n-2}$. The latter subgroup is naturally isomorphic to B_{n-1}, and may so be identified with B_{n-1}. A second application of the Schreier-Reidemeister method treats the subgroup D_{n-1} of B_{n-1}, which is similarly exhibited as a semi-direct product of U_{n-1} and B_{n-2}. Proceeding in this way, one finally obtains that every element of the group B_n is a product of the form $u_2M_{2,k_2}u_3M_{3,k_3}\cdots u_nM_{n,k_n}$ with $u_i \in U_i$, $2 \leq i \leq n$. Each M_{j,k_j} is commutative with u_s if $j < s$, hence every element of the group B_n is of the form $u_2u_3\cdots u_nM$, where $u_2u_3\cdots u_n$ covers all elements of P_n, and where M is a product of the form $M_{2,k_2}M_{3,k_3}\cdots M_{n,k_n}$. The advantage of Chow's method is that the Schreier-Reidemeister calculation need only be applied to a single subgroup of index n, as the calculations for $B_{n-1}, B_{n-2}, \cdots, B_2$ may all be obtained by analogy to the calculation for B_n. ||

We may now establish Theorem 1.8. The group P_{n-1} can be regarded as the subgroup of P_n which is generated by $\{A_{ij},\ 1 \leq i < j \leq n-1\}$. Note that a natural homomorphism $\eta : P_n \to P_{n-1}$ may be defined by the rule $\eta(A_{ij}) = A_{ij}$ if $1 \le i < j \le n-1$, while $\eta(A_{in}) = 1$, $1 \le i < n$. Thus $\ker \eta$ is the normal closure in P_n of the elements $A_{1n}, A_{2n}, \cdots, A_{n-1,n}$. An examination of relations (1-12) shows that in fact the subgroup U_n of P_n which is generated by $A_{1n}, A_{2n}, \cdots, A_{n-1,n}$ is normal in P_n, hence $U_n = \ker \eta$.

Corresponding to the homomorphism $\eta : P_n \to P_{n-1}$ we have the homomorphism $\pi_* : \pi_1 F_{0,n} \to \pi_1 F_{0,n-1}$ of Theorem 1.4. By Theorem 1.4, we also know that $\ker \pi_* = \pi_1 F_{n-1,1} = \pi_1(E^2 - Q_{n-1})$, which is a free group of rank $n-1$.

It is easy to see that the following diagram is commutative:

$$\begin{array}{ccccccccc} 1 & \longrightarrow & U_n & \longrightarrow & P_n & \xrightarrow{\eta} & P_{n-1} & \longrightarrow & 1 \\ & & \downarrow{\scriptstyle \iota_n | U_n} & & \downarrow{\scriptstyle \iota_n} & & \downarrow{\scriptstyle \iota_{n-1}} & & \\ 1 & \longrightarrow & \pi_1 F_{n-1,1} & \longrightarrow & \pi_1 F_{0,n} & \xrightarrow{\pi_*} & \pi_1 F_{0,n-1} & \longrightarrow & 1 \end{array}$$

with exact rows. In the bottom row (following the notation used in Theorem 1.4) the base point for $\pi_1 F_{0,n}$ is $(z_1^0, \cdots, z_n^0)$, so that z_n^0 is the base point for $\pi_1 F_{n-1,1} = \pi_1(E^2 - z_1^0 \cup \cdots \cup z_{n-1}^0)$. Now, from equation (1-11) and our picture definition of the geometric braids $\tilde{\sigma}_i = \iota_n(\sigma_n)$ one may identify the image $\iota_n(A_{jn})$ of the generator A_{jn} of U_n as being represented by a loop based at z_n^0 which encircles the point z_j^0 once and separates it from $z_1^0, \cdots, z_{j-1}^0, z_{j+1}^0, \cdots, z_{n-1}^0$. Clearly the image set $\{\iota_n(A_{jn}); 1 \le j < n\}$ is a free basis for the free group $\pi_1 F_{n-1,1}$. By the Hopfian property for free groups [see Magnus, Karass and Solitar, 1966], it then follows that U_n must also be free and that $\iota_n | U_n$ is an isomorphism onto. Now observe that $P_1 = 1$ and $\pi_1 F_{0,1} = 1$. Therefore ι_1 is an isomorphism. Assume inductively therefore that ι_{n-1} is an isomorphism. Then, since $\iota_n | U_n$ is an isomorphism for each n, ι_n is an isomorphism by the five lemma. This completes the proof of Theorem 1.8. ||

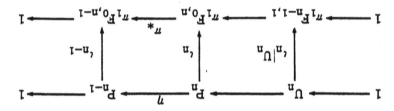

COROLLARY 1.8.1. *The group* P_n *is a semi direct product of* U_n *and* P_{n-1}.

Proof. Implicit in the proof of Theorem 1.8.

DEFINITION. Theorem 1.8 establishes an isomorphism $\iota : B_n \to \pi_1 B_{0,n}$ which takes the abstract group B_n with the presentation of Theorem 1.8 onto the Artin braid group $\pi_1 B_{0,n}$ of the plane E^2. The two groups will now be identified and notation for the two groups used interchangeably. Similarly, the group P_n will be identified with $\pi_1 F_{0,n}$. In particular, elements of B_n (respectively P_n) will be called braids (respectively pure braids) and B_n (respectively P_n) will be called the braid group (pure braid group) of the plane. The coset representatives for P_n in B_n which are defined by equation (1-13) will be called the *permutation braids*. The relations (1-1) and (1-2) will be called the *braid relations*.

COROLLARY 1.8.2. *Every element* $\beta \in B_n$ *can be written uniquely in the form*

$$\beta = \beta_2 \beta_3 \cdots \beta_n \pi_\beta \,, \tag{1-13}$$

where π_β *is a permutation braid and each* β_j *belongs to the free subgroup* U_j *defined in the proof of Theorem 1.8.*

Proof. Since the permutation braids form a complete set of coset representatives for P_n in B_n, $\beta = \overline{\beta}_n \pi_\beta$ for some $\overline{\beta}_n \in P_n$ and permutation braid π_β. By Corollary 1.8.1 and its proof, $\overline{\beta}_n = \overline{\beta}_{n-1} \beta_n$ $(n > 2)$ for some $\beta_n \in U_n$ and $\overline{\beta}_{n-1} \in P_{n-1}$. By induction,

$$\beta = \overline{\beta}_2 \beta_3 \cdots \beta_n \pi_\beta$$

where $\beta_i \in U_i$, $1 = 3, \cdots, n$ and $\overline{\beta}_2 \in P_2 = U_2$. Let $\beta_2 = \overline{\beta}_2$. This completes the proof of existence. The uniqueness of π_β is clear. The uniqueness of each $\beta_i (i = 2, \cdots, n)$ derives, by induction, from the properties of semi-direct products of groups.

Since each β_i of Corollary 1.8.2 lies in a free group on known free generators, it is possible to algorithmically calculate standard representatives for $\beta_2, \beta_3, \cdots, \beta_n$, and π_β in the given generators for B_n. This solves the word problem in B_n. ‖

The procedure for putting a braid word into the normal form (1-13) is called "combing the braid." Note that each entry β_j in (1-13) is a product of the free generators $A_{1j}, \cdots, A_{j-1,j}$ of the free group U_j. The geometric braid corresponding to the generator A_{ij} is illustrated in Figure 4, together with an example of a braid which has been combed. Artin discourages any attempt to carry out this procedure experimentally on a living person, fearing that it would "only lead to violent protests and discrimination against mathematics" [Artin, 1947a, p. 126].

COROLLARY 1.8.3. *The braid group* B_n *has a faithful representation as a group of (right) automorphisms of a free group* $F_n = \langle x_1, \cdots, x_n \rangle$, *of rank* n. *The representation is induced by a mapping* ξ *from* B_n *to* Aut F_n *defined by:*

$$(1\text{-}14) \qquad (\sigma_i)\xi : \begin{cases} x_i \to x_i x_{i+1} x_i^{-1} \\ x_{i+1} \to x_i \\ x_j \to x_j \quad \textit{if} \quad j \neq i, i+1 . \end{cases}$$

The restriction of ξ *to the pure braid group* P_n *maps the generator* A_{rs} *of* P_n *to the automorphism:*

$$(1\text{-}15) \qquad (A_{rs})\xi : x_i \to x_i \quad (\textit{if} \ \ s < i \ \ \textit{or} \ \ i < r)$$

$$\to x_r x_i x_r^{-1} \quad (\textit{if} \ \ s = i)$$

$$\to x_i x_s x_i x_s^{-1} x_i^{-1} \quad (\textit{if} \ \ r = i)$$

$$\to x_r x_s x_r^{-1} x_s^{-1} x_i x_s x_r x_s^{-1} x_r^{-1} \quad (\textit{if} \ \ r < i < s) .$$

Proof. To see that ξ induces a representation is simply a matter of checking that relations are preserved.

In the usual way ξ induces a representation of B_n as a group of automorphisms of the commutator factor group $\dot{F}_n = F_n/[F_n, F_n]$ of F_n. Denoting the generators of $\dot{F}_n$ by $\dot{x}_1, \cdots, \dot{x}_n$, we find from (1-14) that the automorphism of $\dot{F}_n$ induced by σ_i maps $\dot{x}_i \to \dot{x}_{i+1}$, $\dot{x}_{i+1} \to \dot{x}_i$ and $\dot{x}_j \to \dot{x}_j$ if $j \neq i$. Clearly, the automorphism of $\dot{F}_n$ induced by elements of P_n are trivial, and this is a faithful representation of the permutation group $B_n/P_n \simeq \Sigma_n$. Hence it follows from the 5-lemma that ξ will be faithful if $\xi | P_n$ is faithful.

We now show that the representation defined by ξ arises in a natural manner. Recall that $P_{n+1} = P_n \cdot U_{n+1}$. Note that U_{n+1} is a free subgroup of P_{n+1} of rank n. It has already been observed that U_{n+1} is normal in P_{n+1}; hence P_n acts by conjugation as a group of automorphisms of U_{n+1}. Define an isomorphism from U_{n+1} to F_n by sending $A_{j,n+1}$ to x_j for each $j = 1, \cdots, n$. Comparison of equations (1-12) and (1-15) shows that one obtains thereby a commutative diagram

$$\begin{array}{ccc} & \xrightarrow{\text{conjugation}} & \operatorname{Aut} U_{n+1} \\ P_n & & \downarrow \approx \\ & \xrightarrow{\quad \xi \quad} & \operatorname{Aut} F_n \end{array} .$$

Thus kernel ξ is precisely the subgroup of all elements of P_n which commute with U_{n+1}, where both P_n and U_{n+1} are now regarded as elements of P_{n+1}.

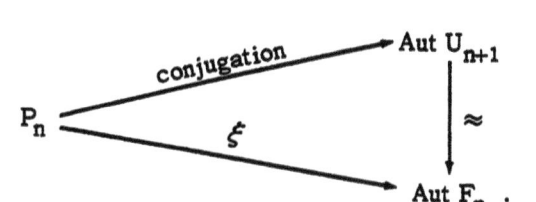

Suppose, now, that $\beta \in \ker \xi$, with $\beta \neq 1$. By Corollary 1.8.2 we may then write β in the form $\beta = \beta_2 \cdots \beta_{i-1} \beta_i$, where i is the largest integer such that $\beta_i \neq 1$, but $\beta_{i+1} = \beta_{i+2} = \cdots = \beta_n = 1$. Now, β commutes with each element of U_{n+1}, hence β commutes with $A_{i,n+1}$. By equation (1-11), the element $A_{i,n+1}$ depends only on $\sigma_i, \cdots, \sigma_n$. Note that each β_j $(2 \leq j \leq i)$ belongs to the free subgroup U_j of P_{n+1} freely generated by $A_{1,j}, \cdots, A_{j-1,j}$, and hence (by equation (1-11)) β_j depends only on $\sigma_1, \cdots, \sigma_{j-1}$. Therefore, as a consequence of relation (1-1) it follows that β_j commutes with $A_{i,n+1}$ whenever $j \leq i-1$. Therefore, the condition that β commutes with $A_{i,n+1}$ implies:

$$\beta_i A_{i,n+1} \beta_i^{-1} = A_{i,n+1} \quad (i = 1, \cdots, n) . \tag{1-16}$$

We will now show that (1-16) implies that β_i is 1, giving the sought-for contradiction, so that kernel $\xi = 1$.

To motivate the algebraic manipulations which follow, we remark that the elements $\{A_{1,i}, \cdots, A_{i-1,i}, A_{i,i+1}, \cdots, A_{i,n+1}\}$ generate a free subgroup of P_{n+1} which is naturally isomorphic to the subgroup U_{n+1} freely generated by $\{A_{1,n+1}, A_{2,n+1}, \cdots, A_{n,n+1}\}$, for each $i = 2, \cdots, n-1$. This is intuitively clear because it is quite arbitrary how we assign indices to the braid "strings." We will now establish this fact algebraically, in order to be able to use the fact that equation (1-16) is a statement that a pair of elements in a free group commute, and hence to conclude that β_i and $A_{i,n+1}$ are powers of the same element.

Let $\pi = \sigma_n \sigma_{n-1} \cdots \sigma_i$. Using relations (1-1) and (1-2) and equation (1-11) we claim that

(1-17)[3] $$\pi A_{i,n+1} \pi^{-1} = A_{n,n+1} \quad (i = 1, \cdots, n)$$

(1-18)[3] $$\pi A_{k,i} \pi^{-1} = A_{k,n+1} \quad (k = 1, \cdots, i-1) .$$

---

3 To establish equation (1-17), note first that $A_{i,n+1} = \sigma_n \cdots \sigma_{i+1} \sigma_i^2 \sigma_{i+1}^{-1} \cdots \sigma_n^{-1} = \sigma_i^{-1} \cdots \sigma_{n-1}^{-1} \sigma_n^2 \sigma_{n-1} \cdots \sigma_i$. The easiest way to see this is to inspect the geometric braid $A_{i,n+1}$ (Figure 4), and to observe that when the $n+1^{st}$ string is pulled taut, with the ith string loose, the geometric braid defined by the left expression for $A_{i,n+1}$ goes over to the geometric braid defined by the expression on the right. This can also be established algebraically, as a consequence of equations (1-1) and (1-2). Using the expression on the right above for $A_{i,n+1}$, equation (1-17) then follows easily. Equation (1-18) is an immediate consequence of the definitions of $A_{k,i}$ and of $A_{k,n+1}$ as given by equation (1-11).

Transforming equation (1-16) by π, and applying (1-17), we obtain:

$$(\pi\beta_i\pi^{-1})A_{n,n+1}(\pi\beta_i^{-1}\pi^{-1}) = A_{n,n+1}\,. \tag{1-19}$$

But, by equations (1-18), the element $\pi\beta_i\pi^{-1}$ belongs to the free group U_{n+1}, and if two elements in a free group commute, then they must each be powers of some element in that group. Since $A_{n,n+1}$ is a generator of U_{n+1}, it then follows that $\pi\beta_i\pi^{-1} = A^s_{n,n+1}$ for some integer s. But then $\beta_i = \pi^{-1}A^s_{n,n+1}\pi$, and by equation (1-17) this is precisely $A^s_{i,n+1}$. We now have the sought-for contradiction, for β_i belongs to the free group U_i, and the only way that $A^s_{i,n+1}$ can be in U_i is if $s = 0$, giving $\beta_i = 1$. But then $\beta = 1$, hence $\ker \xi = 1$. ||

From now on we will use the symbol B_n to mean not only the abstract group of Theorem 1.8, and the geometric braid group $\pi_1 B_{0,n}E^2$, but also its realization as a group of right automorphisms of F_n. That is, we will replace the symbols $(B_n)\xi$, $(P_n)\xi$, $(U_n)\xi$, $(\sigma_i)\xi$, $(A_{ij})\xi$ by B_n, P_n, U_n, σ_i, A_{ij} respectively.

COROLLARY 1.8.4. *If* $n \geq 3$ *the center of* B_n *is the infinite cyclic subgroup generated by* $(\sigma_1\sigma_2\cdots\sigma_{n-1})^n = (A_{12})(A_{13}A_{23})\cdots(A_{1n}A_{2n}\cdots A_{n-1,n})$. [Chow, 1948].

Proof. We leave it to the reader to verify that as a consequence of relations (1-12):

(i) The element $(A_{12})(A_{13}A_{23})\cdots(A_{1n}A_{2n}\cdots A_{n-1,n}) \in$ Center P_n.

(ii) The element $(A_{1n}A_{2n}\cdots A_{n-1,n}) \in$ Centralizer of P_{n-1} in P_n, where P_{n-1} is regarded as the subgroup of P_n which is generated by $\{A_{rs};\ 1 \leq r < s \leq n-1\}$.

Suppose that $\beta \in$ Center B_n. The symmetric group Σ_n is centerless for $n \geq 3$, hence β must be in the kernel of the homomorphism $\nu : B_n \to \Sigma_n$, i.e., $\beta \in P_n$. By Corollary 1.8.1 it then follows that β has a unique representation $\beta = \bar{\beta}_{n-1}\beta_n$ where $\beta_n \in U_n$, $\bar{\beta}_{n-1} \in P_{n-1}$. The condition that $\beta \in$ Center B_n then implies:

$$\bar{\beta}_{n-1}\beta_n A_{in}\beta_n^{-1}\bar{\beta}_{n-1}^{-1} = A_{in} \quad 1 \le i \le n-1 . \tag{1-20}$$

Multiplying together the $n-1$ equations obtained by setting $i = 1,2,\cdots,n-1$ in equation (1-20), we get

$$\beta_n(A_{1n}A_{2n}\cdots A_{n-1,n})\beta_n^{-1} = \bar{\beta}_{n-1}^{-1}(A_{1n}A_{2n}\cdots A_{n-1,n})\bar{\beta}_{n-1} .$$

Since $\bar{\beta}_{n-1} \in P_{n-1}$, condition (ii) above then implies:

$$\beta_n(A_{1n}A_{2n}\cdots A_{n-1,n})\beta_n^{-1} = (A_{1n}A_{2n}\cdots A_{n-1,n}) .$$

This equality holds in the free group U_n, hence the only possibility is

$$\beta_n = (A_{1n}A_{2n}\cdots A_{n-1,n})^m$$

for some integer m. Using this information in (1-20), we thus obtain:

$$\bar{\beta}_{n-1}^{-1}A_{in}\bar{\beta}_{n-1} = (A_{1n}A_{2n}\cdots A_{n-1,n})^m A_{in}(A_{1n}A_{2n}\cdots A_{n-1,n})^{-m}. \tag{1-21}$$

Equation (1-21) expresses the action of the element $\bar{\beta}_{n-1}$ on the free generators $A_{1n}, A_{2n},\cdots,A_{n-1,n}$ of U_n by conjugation. By Corollary 1.8.3, this action induces a faithful representation of P_{n-1} as a group of automorphisms of the free group U_n. Now, a calculation based on equations (1-12) shows that the element $[(A_{12})(A_{13}A_{23})\cdots(A_{1,n-1}A_{2,n-1}\cdots A_{n-2,n-1})]^{-m}$ has precisely the effect which our element $\bar{\beta}_{n-1}^{-1}$ is required to have, hence $\bar{\beta}_{n-1}^{-1}$ must be precisely $[(A_{12})(A_{13}A_{23})\cdots(A_{1,n-1}A_{2,n-1}\cdots A_{n-2,n-1})]^{-m}$ for some integer m. Thus our original element $\beta = \bar{\beta}_{n-1}\beta_n$ can only have been:

$$\beta = [(A_{12})(A_{13}A_{23})\cdots(A_{1,n-1}A_{2,n-1}\cdots A_{n-2,n-1})(A_{1,n}A_{2n}\cdots A_{n-1,n})]^m.$$

(In this last step we made use of property (ii) above.) Since by property (i) this element is in fact in the center for *every* integer m, the proof of Corollary 1.8.4 is complete. ‖

THEOREM 1.9 [Artin, 1925]. *Let* $F_n = \langle x_1, \cdots, x_n \rangle$ *be a free group of rank* n. *Let* β *be an endomorphism of* F_n. *Then* $\beta \in B_n \subset \operatorname{Aut} F_n$ *if and only if* β *satisfies the two conditions*

$$(x_i)\beta = A_i x_{\mu_i} A_i^{-1} \qquad 1 \le i \le n \tag{1-22}$$

$$(x_1 x_2 \cdots x_n)\beta = x_1 x_2 \cdots x_n \tag{1-23}$$

where $(\mu_1, \cdots, \mu_n)$ *is a permutation of* $(1, \cdots, n)$, *and* $A_i = A_i(x_1, \cdots, x_n)$ *is a word in the generators of* F_n.

Proof. The necessity of conditions (1-22) and (1-23) are immediate, therefore we need only establish that they are sufficient. This will be accomplished by proving that every endomorphism of F_n which satisfies (1-22) and (1-23) is a product of powers of $\sigma_1, \cdots, \sigma_{n-1}$, and hence is in B_n. To prove this, we examine how cancellations occur in the equality

$$A_1 x_{\mu_1} A_1^{-1} A_2 x_{\mu_2} A_2^{-1} \cdots A_n x_{\mu_n} A_n^{-1} = x_1 x_2 \cdots x_n \tag{1-24}$$

which results from (1-22) and (1-23). It is assumed that each term $A_i x_{\mu_i} A_i^{-1}$ is freely reduced.

We assert that in order for (1-24) to hold in the free group F_n, there must exist some $\nu = 1, \cdots, n-1$ such that either

(a) $x_{\mu_\nu} A_\nu^{-1}$ is absorbed by $A_{\nu+1}$

or

(b) A_ν^{-1} absorbs $A_{\nu+1} x_{\mu_{\nu+1}}$.

This assertion will imply the truth of Theorem 1.9 by the following reasoning: Define the "length" of the automorphism β to be the sum of the letter lengths of the words $A_i x_{\mu_i} A_i^{-1}$, $1 \le i \le n$. We will show that if assertion (a) is true, then $\sigma_\nu \beta$ has shorter length than β, while if (b) is true then $\sigma_\nu^{-1}\beta$ has shorter length than β. (Remember that braid automorphisms act on the right, hence $\sigma_\nu \beta$ means apply σ_ν first, then apply β.) This implies that every automorphism β of F_n which satisfies conditions (1-22) and (1-23) can be reduced to the identity by repeated application of appropriate elementary automorphisms σ_ν or σ_ν^{-1}. Thus β is a power product of the σ_i, hence $\beta \in B_n$.

To show that the length can always be reduced as indicated, suppose first that (a) is true. Then the action of β on x_ν and $x_{\nu+1}$ is given by:

$$(1\text{-}25) \qquad (x_\nu)\beta = A_\nu x_{\mu_\nu} A_\nu^{-1}$$

$$(x_{\nu+1})\beta = A_\nu x_{\mu_\nu}^{-1} \tilde{A}_{\nu+1} x_{\mu_{\nu+1}} \tilde{A}_{\nu+1}^{-1} x_{\mu_\nu} A_\nu^{-1} .$$

Using the action given in (1-14), we now compute the product $\sigma_\nu \beta$:

$$(1\text{-}26) \qquad (x_\nu)\sigma_\nu \beta = A_\nu \tilde{A}_{\nu+1} x_{\mu_{\nu+1}} \tilde{A}_{\nu+1}^{-1} A_\nu^{-1}$$

$$(x_{\nu+1})\sigma_\nu \beta = A_\nu x_{\mu_\nu} A_\nu^{-1} .$$

Since both β and $\sigma_\nu \beta$ have the same effect on x_j if $j \ne \nu, \nu+1$, a comparison of (1-25) and (1-26) shows that $\sigma_\nu \beta$ has shorter length than β. An identical argument holds in case (b). Thus Theorem 1.9 will be true if we can show that our assertion about cancellation is true.

We examine the manner in which the left hand side of (1-24) reduces to the right hand side. Suppose first that one of the terms $A_\nu x_{\mu_\nu} A_\nu^{-1}$ is completely absorbed by the other terms in the free cancellations which reduce the LHS of (1-24) to the RHS. We ask how the letter x_{μ_ν} is absorbed in these cancellations? If x_{μ_ν} is absorbed by a letter to the left of $x_{\mu_{\nu-1}}$, then (a) is satisfied. If x_{μ_ν} is absorbed by a letter in $A_{\nu-1}^{-1}$, then (b) is satisfied. If x_{μ_ν} is absorbed by a letter in $A_{\nu+1}$, then (a) is satisfied. If x_{μ_ν} is absorbed by a letter to the right of $x_{\mu_{\nu+1}}$, then (b) is satisfied. Since x_{μ_ν} cannot be absorbed by either $x_{\mu_{\nu-1}}$ or $x_{\mu_{\nu+1}}$, both of which have subscripts which are different from the subscript μ_ν, all possible cases have been treated.

It remains to consider the case where there is *no* subscript ν with the property that $A_\nu x_{\mu_\nu} A_\nu^{-1}$ is completely absorbed. In this case, some residue R_ν will remain for each $A_\nu x_\mu A_\nu^{-1}$ after all free reductions have been made. Then (1-24) implies $R_1 R_2 \cdots R_n = x_1 x_2 \cdots x_n$. This implies that $R_i = x_i$ for each $i = i, \cdots, n$. Now examine the term $A_1 x_{\mu_1} A_1^{-1}$. The initial letter in this term can only be x_1. We consider first the case where $A_1 x_{\mu_1} A_1^{-1}$ is not identically x_1, i.e.,

$$A_1 x_{\mu_1} A_1^{-1} = x_1 \bar{A}_1 x_{\mu_1} \bar{A}_1^{-1} x_1^{-1} .$$

Since by hypothesis $\bar{A}_1 x_{\mu_1} \bar{A}_1^{-1} x_1^{-1}$ is completely absorbed, there are two possibilities: x_{μ_1} is absorbed by A_2 (in which case (a) is satisfied) or by a letter to the right of x_{μ_2} (in which case (b) is satisfied). If, on the other hand, $A_1 x_{\mu_1} A_1^{-1} = x_1$ the entire argument can be repeated for $A_2 x_{\mu_2} A_2^{-1}$, etc. In this way we see that in every case either (a) or (b) is true, hence Theorem 1.9 is established. ||

Using Theorem 1.9, we will now be able to place a new geometric interpretation on the group B_n.

Let D^2 be a disc, and let $Q_n = \{q_1, \cdots, q_n\}$ be a set of fixed, distinguished points of D^2. The fundamental group $\pi_1(D^2 - Q_n)$ is a free group of rank n. Let $x_1, \cdots, x_n$ be a basis for $\pi_1(D^2 - Q_n)$, where $x_i (i = 1, \cdots, n)$ is represented by a simple loop which encloses the boundary point q_i, but no boundary point q_j for $j \neq i$. See Figure 5.

THEOREM 1.10. *Let* M *be the group of automorphisms of* $\pi_1(D^2 - Q_n)$ *which are induced by homeomorphisms of* $D^2 - Q_n$ *which keep the boundary of* D^2 *fixed pointwise. Then* M *is precisely the group* B_n.

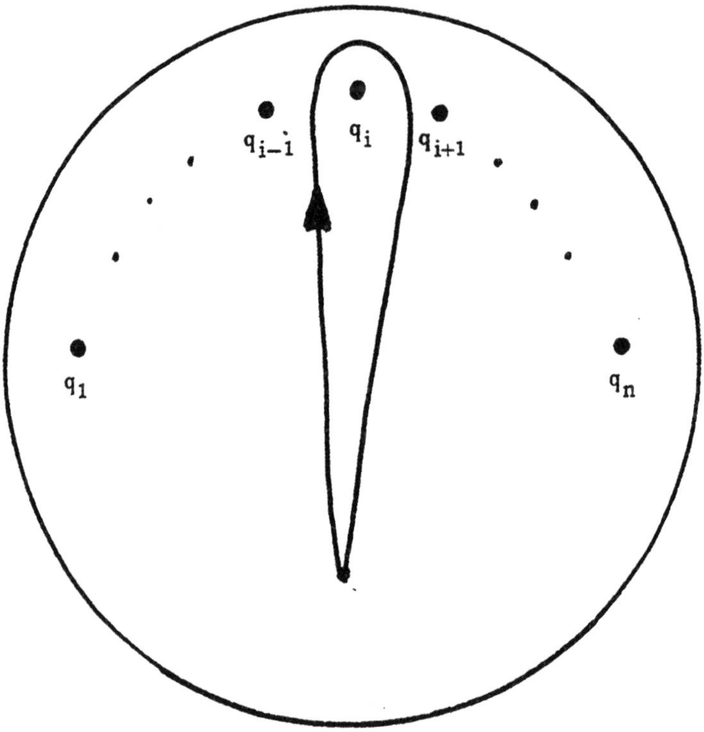

Fig. 5.

Proof of Theorem 1.10. For geometric reasons, each element $\beta \in M$ satisfies conditions (1-22) and (1-23), hence $M \subseteq B_n$.

Recall that B_n is generated by the automorphisms $\sigma_1, \cdots, \sigma_{n-1}$ (cf. equation (1-14)). Since the action of σ_i on F_n can be realized by a homeomorphism which maps the point $q_i \to q_{i+1}$, maps $q_{i+1} \to q_i$, and is the identity outside a disc which includes q_i and q_{i+1} (but no q_j for $j \neq i, i+1$) it follows that $B_n \subseteq M$. Hence $B_n \equiv M$. ‖

Remark. Let Δ be an autohomeomorphism of $D^2 - Q_m$ which keeps the boundary of D^2 fixed pointwise. Thus Δ represents an element of M. Then Δ has a unique extension $\bar{\Delta}$ to D^2 which permutes the points of Q_m. The map $\bar{\Delta}$ is isotopic to the identity in D^2. This isotopy may be used to define an autohomeomorphism of $D^2 \times I$ which preserves $D^2 \times \{0\}$ pointwise, preserves $D^2 \times \{t\}$, $t \in I$, setwise, and coincides with $\bar{\Delta}$ on $D^2 \times \{1\}$. The image of $Q_n \times I$ under this extension is a geometric braid in the sense defined earlier.

1.5. *Braid group of the sphere*

THEOREM 1.11 [Fadell-Van Buskirk, 1962]. *The braid group* $\pi_1 B_{0,n} S^2$ *of the 2-sphere* S^2 *admits a presentation with generators* $\delta_1, \cdots, \delta_{n-1}$ *and defining relations*:

(1-27) $$\delta_i \delta_j = \delta_j \delta_i \quad \textit{if} \quad |i-j| \geq 2, \ i \leq i,j \leq n-1$$

(1-28) $$\delta_i \delta_{i+1} \delta_i = \delta_{i+1} \delta_i \delta_{i+1} \quad 1 \leq i \leq n-2$$

(1-29) $$\delta_1 \cdots \delta_{n-2} \delta_{n-1}^2 \delta_{n-2} \cdots \delta_1 = 1 .$$

Proof. The proof will only be outlined. It rests on an inductive argument which is exactly like that used in the proof of Theorem 1.8. The only difficulty is that the fundamental exact sequence (1-7), which was crucial in the establishment of Theorem 1.8, is only valid for $n \geq 4$ when $M = S^2$. Therefore the inductive argument which was the basis of the proof of Theorem 1.8 begins with $n = 3$ rather than 1. The following additional facts are needed to establish Theorem 1.11:

(i) $\pi_1 F_{0,2} S^2 = 1$

(ii) $\pi_1 B_{0,2} S^2 =$ cyclic group of order 2

(iii) $\pi_1 F_{0,3} S^2 =$ cyclic group of order 2

(iv) $\pi_1 B_{0,3} S^2 =$ ZS metacyclic group of order 12

(v) $\pi_2 F_{0,3} S^2 = 1$.

The first of these follows easily from the fibration of Theorem 1.2 using the well-known facts that $\pi_1 F_{1,1} S^2$ and $\pi_1 F_{0,1} S^2$ are both trivial groups.

To prove (ii), one need only note that $\pi_1 B_{0,2}S^2$ maps homomorphically onto Σ_2, which is of order 2, and use (i). For proofs of (iii)-(v) the reader is referred to Fadell and Van Buskirk, 1967. ||

We note, for completeness, that the faithful representation found for $\pi_1 B_{0,n}E^2$ as a group of automorphisms of the fundamental group of the n-punctured plane (Corollary 1.8.3 and Theorem 1.10) does not generalize to a faithful representation of $\pi_1 B_{0,n}S^2$ as a group of automorphisms of the fundamental group of the n-punctured sphere. To be sure, the action given in equations (1-14) does induce an action of $\pi_1 B_{0,n}S^2$ on $\hat{F}_{n-1}$, where $\hat{F}_{n-1}$ is the quotient group of F_n obtained by adding the single relation $x_1 x_2 \cdots x_n = 1$. This action is, however, not a representation of $\pi_1 B_{0,n}S^2$, because relation (1-29) is not satisfied.[4] Moreover, one may verify without difficulty that the element $(\delta_1 \delta_2 \cdots \delta_{n-1})^n$ induces the identity automorphism of $\hat{F}_{n-1}$, yet the relation $(\delta_1 \delta_2 \cdots \delta_{n-1})^n$ cannot be a consequence of relations (1-27), (1-28) and (1-29).[5] This induced action on $\hat{F}_{n-1}$ will be studied further in Chapter 4, Section 4.2.

1.6. *Survey of 2-manifold braid groups.* If M is a closed orientable 2-manifolds and either $n \geq 4$ or $P^2 \neq M \neq S^2$ and $n \geq 2$, then as shown in Theorem 1.4, there is an exact sequence

$$1 \to \pi_1 F_{n-1,1}M \to \pi_1 F_{0,n}M \to \pi_1 F_{0,n-1}M \to 1 .$$

This sequence was the basis for the structural analysis of $\pi_1 B_{0,n}E^2$ carried out in Section 1.3. The same sort of analysis can be carried out with successively more difficulty for other 2-manifold braid groups. The results appear in the following papers:

---

4 $\delta_1 \cdots \delta_{n-2} \delta_{n-1}^2 \delta_{n-2} \cdots \delta_1$ induces an inner automorphism of $\hat{F}_{n-1}$, and the normal closure of this element is the full group $\mathrm{Inn}\, \hat{F}_{n-1}$.

5 See [Fadell and Van Buskirk, 1962].

E^2[Chow (1948); Fadell-Van Buskirk (1962)]

P^2[Van Buskirk (1966)]

S^2 [Fadell-Van Buskirk (1962)]

Torus [Birman (1969a)]

All closed 2-manifolds [G. P. Scott (1970)].

CHAPTER 2
BRAIDS AND LINKS

Our focus in this chapter will be on Artin's braid group $\pi_1 B_{0,n}E^2$, which as before will be denoted simply by B_n. In particular, we will be interested in the possibility of using braid theory as an approach to the study of knots and links.

We will begin by defining the notion of a closed braid, and proving the basic result, due to Alexander, that every link type may be represented by a closed braid (Theorem 2.1). We will then establish a second classical result (Theorem 2.2) due to Artin,[1] which establishes that braid automorphisms may be used to obtain presentations for the fundamental group of the complement of any tame link in S^3, and moreover that braid automorphisms may be used to give a complete characterization of link groups as a class of groups which admit certain canonical presentations (Corollary 2.2.1).

Section 2.2 is devoted to a result of Markov (Theorem 2.3). The statement of Markov's theorem begins with a listing of certain special moves, denoted type $\mathfrak{R}$ and type $\mathfrak{W}$, which may be applied to links, and which take closed braids to closed braids, and do not alter link type. Markov's theorem establishes that a finite sequence of these moves suffices to take any closed braid representative V of a given link type to any other closed braid representative V' of the same link type. This theorem is of particular interest because it allows us to restate the link problem as a purely

1 Our proof of Theorem 2.2 is, however, different from Artin's proof.

algebraic problem (which we denote "the algebraic link problem") about the sequence of braid groups $B_1, B_2, B_3, \cdots$. This restatement will be accomplished in Corollary 2.3.1.

The "algebraic link problem," as defined in Corollary 2.3.1, includes as a sub-problem the conjugacy problem in the braid group B_n. For this reason, it is of particular interest that an algorithm now exists to solve the conjugacy problem in B_n. This algorithm, which is due to F. Garside, will be presented in Theorem 6 of Section 2.3. The algorithm is, unfortunately, quite complicated, as well as difficult to establish, and it is clear that further work is required before it can be used profitably to attack the algebraic link problem. As a first step in this direction, we present a new result of the author (Theorem 2.7) which leads to a simplified solution to the conjugacy problem (Corollary 2.7.1), and also gives some insight into the properties of a braid word which cause it to be in one conjugacy class rather than another.

At the conclusion of this chapter, in Section 2.4, we will attempt to indicate some of the difficulties and problems encountered when one attempts to apply braid theory to link theory.

A semi-linear viewpoint will be adopted in this chapter. Thus all curves will be considered to be polygonal, and all isotopies will be assumed to be piecewise linear.

2.1. *Closed braids and links*

DEFINITION. A *link* V is the union of $\mu \geq 1$ mutually disjoint simple closed polygonal curves, embedded in E^3. The case $\mu = 1$ will sometimes be referred to as a *knot*.

NOTATION. If $a_1, a_2, \cdots$ are points in E^3, then $[a_1, a_2, \cdots, a_n]$ denotes the convex hull of the points $a_1, a_2, \cdots, a_n$. The symbol ab will sometimes also be used for $[a, b]$. The symbol $[a_1, \cdots, a_n]V$ means

$[a_1,\cdots,a_n] \cap V$. The symbol $[a_1,\cdots,a_n][b_1,\cdots,b_n]$ means $[a_1,\cdots,a_n] \cap [b_1,\cdots,b_n]$.

DEFINITION. Let V be a link with edge ac, and let b be a point which is not on V. Suppose that

$$[a][b,c] = [a,b][c] = \emptyset$$

$$[a,b,c]V = [a,c] .$$

Then we define

$$\mathfrak{E}^b_{ac}V = V - ac + ab + bc \tag{2-1}$$

and say that $\mathfrak{E}^b_{ac}$ *is applicable to* V. The operation $\mathfrak{E}^b_{ac}$ and its inverse are called *elementary* or *type* $\mathfrak{E}$ deformations. The operation $\mathfrak{E}^b_{ac}$ alters V by the adjunction of a new vertex, b, while the operation $(\mathfrak{E}^b_{ac})^{-1}$ removes the vertex b.

Two links V, V′ are said to be *combinatorially equivalent* if there exists a finite sequence of links joining V to V′ with the property that each link in the sequence can be obtained from its predecessor by a single type $\mathfrak{E}$ deformation. The combinatorial equivalence class of a link V will be referred to as a *link isotopy type* [cf. Crowell and Fox, 1963, p. 8]. Let ℓ be an arbitrary, but henceforth fixed, line in E^3 which does not meet the link V. The line ℓ will be referred to as the *axis*. V will be said to be in *general position* with respect to ℓ if none of its edges are coplanar with ℓ.

LEMMA 2.1.1. *Every link is combinatorially equivalent to some link in general position.*

Proof. If V contains an edge ac which is coplanar with ℓ, choose a point b not on V and not in the plane of ℓ and ac, but very close to the edge ac. Then $\mathfrak{E}^b_{ac}V$ has one less edge which is coplanar with ℓ. Continuing this way, we may remove all edges which are coplanar with ℓ. ||

We next define a special type of link which is known as a "closed braid." We first assign an orientation to V. This induces an orientation in each edge. If V is in general position with respect to ℓ, then V may be conveniently pictured by means of its projection onto a plane E^2 which is perpendicular to ℓ at $\hat{\ell}$ (see Figure 4). We now assign a positive direction of rotation about ℓ, which will be indicated on the link diagram by a small arrow $\circlearrowleft$ about the point $\hat{\ell}$ where ℓ pierces E^2. Let ab be an (oriented) edge of V. The edge ab will be said to be *positive* (respectively *negative*), denoted $ab > 0$ (respectively $ab < 0$) if a radius vector from $\hat{\ell}$ to ab rotates in a positive (respectively negative) direction about ℓ in going from a to b along ab. Note that, since V is in general position, every edge of V is either positive or negative. A link is said to be a *closed braid*[2] if all of its edges are positive. The ***height*** of a link, denoted h(V), is the number of negative edges, and is a measure of how far the link is from being a closed braid.

To illustrate these ideas, see Figure 6. The links V and V′ both represent the trefoil knot type. The edges cd, de, ef and fg of the link V are negative, hence V has height 4; hence V is not a closed braid. The link V′ has height 0, and hence is a closed braid. A radius vector from $\hat{\ell}$ to the link V′ never ceases to rotate in a positive direction about ℓ as the link V′ is traversed in the direction of the arrow.

We note that any geometric braid (or *open braid*) β may be used to construct a closed braid $\hat{\beta}$, by identifying the initial points and end points of each of the braid strings, (cf. Chapter 1). Also, that an open braid representative of a given closed braid may be obtained by cutting open the closed braid at its points of intersection with a plane through the axis; the braid word corresponding to β may then be read off from a projection of $\hat{\beta}$.

---

2 We stress that a closed braid is an *oriented* link.

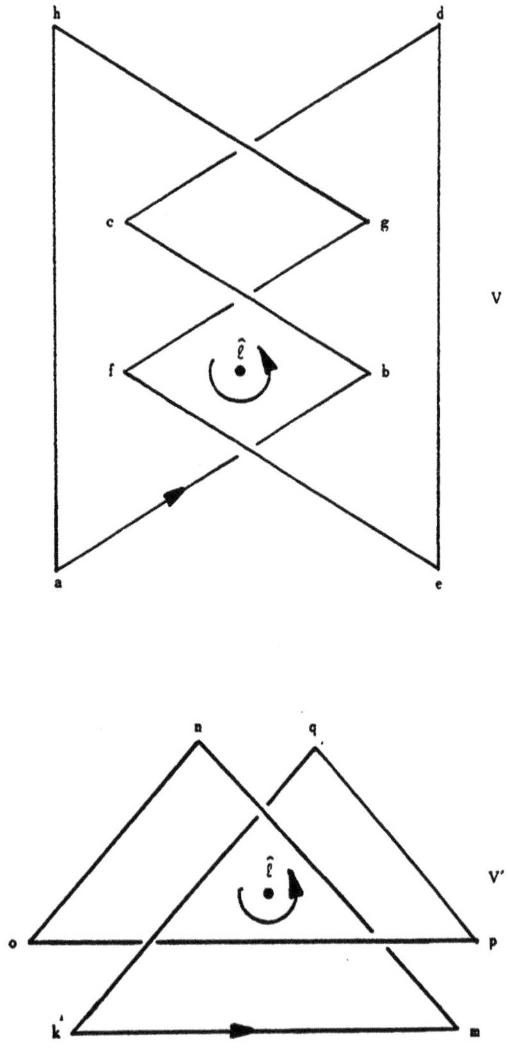

Fig. 6.

Closed braids seem, at first glance, to constitute a very special class of links. The following basic theorem, due to Alexander, will be of fundamental importance in what follows:

THEOREM 2.1 [Alexander, 1923a]. *Every link is combinatorially equivalent to a closed braid.*

Proof. For the remainder of this chapter, it will be assumed that all links are oriented. The proof begins with a definition and a lemma.

DEFINITION. Let $a_0, \cdots, a_m$, $b_0, \cdots, b_m$ be given points such that $a_{i-1}b_i > 0$, $b_i a_i > 0$, $a_{i-1}a_i < 0$, and $a_i \in [a_0, a_m]$, $i = 1, \cdots, m$. Then if

$$\sum_{i=1}^{m} [a_{i-1}, b_i, a_i] V = [a_0, a_m] ,$$

we set

$$\text{(2-2)} \qquad \mathcal{S}^{b_1 \cdots b_m}_{a_0 \cdots a_m} V = \left(\prod_{i=m}^{1} \mathcal{E}^{b_i}_{a_{i-1}, a_i} \right) \left(\prod_{i=m-1}^{1} \mathcal{E}^{a_i}_{a_{i-1}, a_m} \right) V$$

$$= V - a_0 a_n + \sum_{i=1}^{n} (a_{i-1} b_i + b_i a_i) .$$

We call $\mathcal{S}^{b_1 \cdots b_m}_{a_0 \cdots a_m}$ an operation of *type* $\mathcal{S}$. If $\mathcal{S}^{b_1 \cdots b_m}_{a_0 \cdots a_m}$ is applicable to V we say that the set of lines $a_{i-1}b_i$ and $b_i a_i$ is a *sawtooth* on $a_0 a_n$. The effect of applying a type $\mathcal{S}$ operation is to replace a single negative edge with a sequence of positive edges.

The reader is referred to Figure 7(a) for a picture of a sawtooth. In the case illustrated, $m = 3$, and the operation $\mathcal{S}^{b_1 b_2 b_3}_{a_0 a_1 a_2 a_3}$ is applicable to V. The effect of applying this operation is to replace the negative edge $a_0 a$ of the link by the sequence of positive edges $a_0 b_1 + b_1 a_1 + a_1 b_2 + b_2 a_2 + a_2 b_3 + b_3 a$. This is accomplished by first applying the operations $\mathcal{E}^{a_2}_{a_1 a} \mathcal{E}^{a_1}_{a_0 a}$ to V (reading from right to left), in order to subdivide the negative edge $a_0 a$ into a suitable number of subedges. The operations $\mathcal{E}^{b_3}_{a_2 a_3}, \mathcal{E}^{b_2}_{a_1 a_2}, \mathcal{E}^{b_1}_{a_0 a_1}$ are then applied to replace the new negative edges $a_0 a_1, a_1 a_2, a_2 a_3$ by appropriate positive edges. Note that if the link V has height h, the link $\mathcal{S}^{b_1 \cdots b_m}_{a_0 \cdots a_m} V$ will always have height $h - 1$, independent of the integer m.

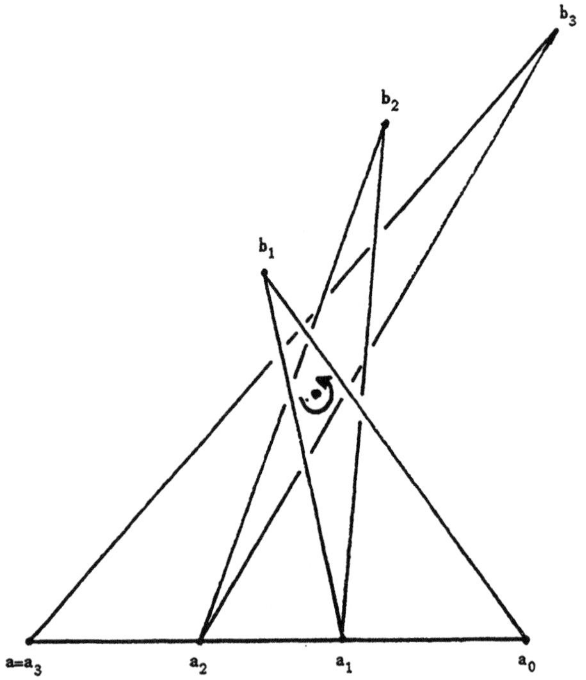

Fig. 7.

(a) A sawtooth on a_0a.

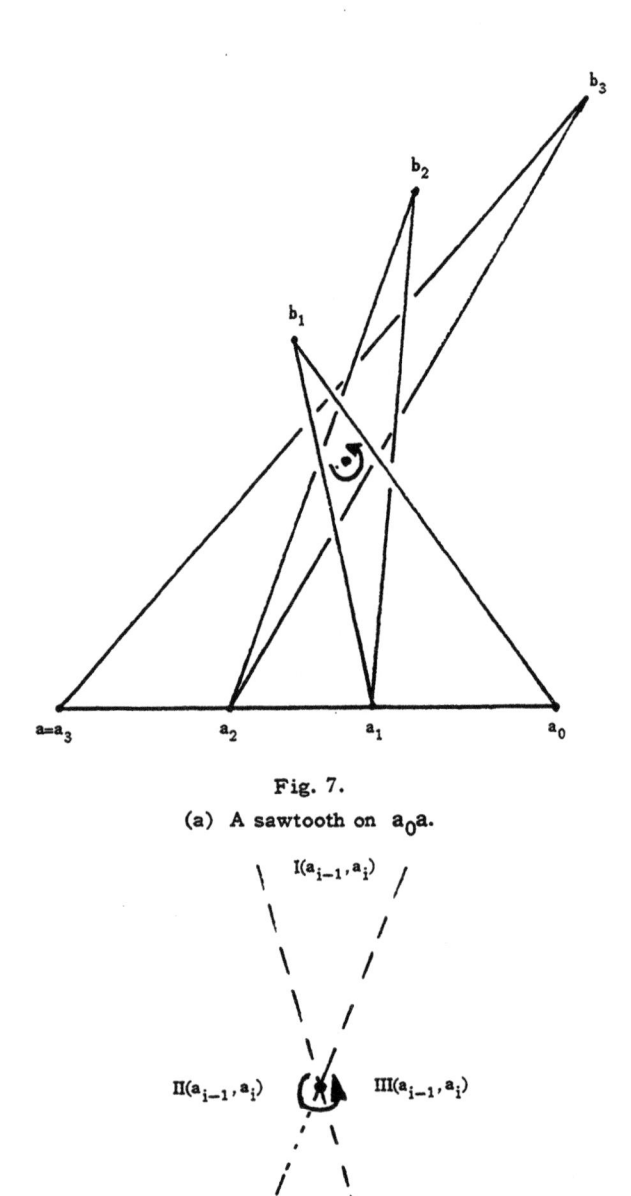

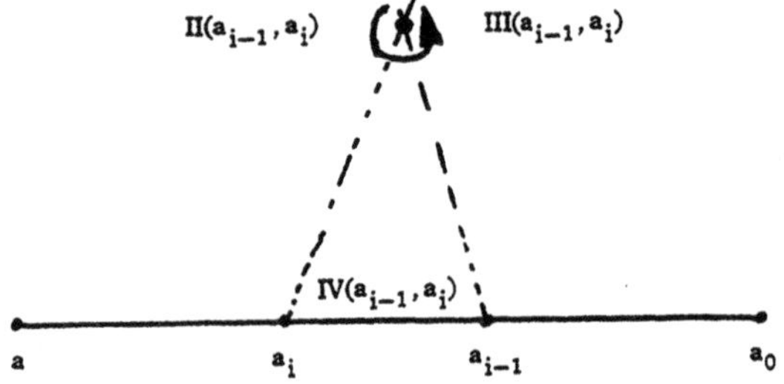

(b) Locating the tip of a sawtooth.

If S is a simplex in 3-space, then a sawtooth will be said to *avoid* S if

$$\bigcup_{i=1}^{m} [a_{i-1}, b_i, a_i] \cap S = \emptyset \text{ or } [a_0] \text{ or } [a_m].$$

LEMMA 2.1.2.[3] *Let* V *be a link which is in general position, and let* $[a_0, a]$ *be any negative edge of* V. *Then a sawtooth may be erected on* $[a_0, a]$. *Moreover, suppose* S *is a simplex in 3-space which has the properties:*

(i) $S \cap [a_0, a_1] = \emptyset$ *or* $[a_0]$ *or* $[a_1]$,

(ii) *If* $S \cap [a_0, a_1] \neq \emptyset$, *then* $S \cap [a_0, a_1]$ *is a vertex of* S.

(iii) *The 1-simplices of* S *are not parallel to* ℓ.

Then the sawtooth may be chosen to avoid S.

Before proving Lemma 2.1.2, we show that it implies the truth of Theorem 2.1. Let V be an arbitrary link, and suppose that $h(V) = h$. If $h = 0$, then V is a closed braid and we are done. If $h > 0$, choose any negative edge a_0a of V. By Lemma 2.1.2, we may find points $a_0, a_1, a_2, \cdots, a_m = a$ and $b_1, b_2, \cdots, b_m$ such that $\mathcal{S}^{b_1 \cdots b_m}_{a_0 \cdots a_m}$ is applicable to V. Then $\mathcal{S}^{b_1 \cdots b_m}_{a_0 \cdots a_m} V$ contains one less negative edge than V, and is combinatorially equivalent to V. Induction on h completes the proof. ||

Proof of Lemma 2.1.2. We wish to show that a sawtooth may be erected on a_0a which avoids S. We will first show that such a sawtooth may always be erected, and then show that it may be modified if necessary to avoid S.

If a_{i-1}, a_i are any points on a_0a such that $a_{i-1}, a_i < 0$, then the planes through the axis ℓ and the points a_{i-1} and a_i respectively divide 3-space into 4-regions, which we label $I(a_{i-1}, a_i)$, $II(a_{i-1}, a_i)$,

---

[3] Lemma 2.1.2 is a somewhat more general result then is needed for the proof of Theorem 2.1. However, we will require the stronger result later, hence we establish it now to avoid repetition.

$III(a_{i-1}, a_i)$ and $IV(a_{i-1}, a_i)$, as in Figure 7. Observe that if b_i is any point in the region $I(a_{i-1}, a_i)$ then we will have $a_{i-1}b_i > 0$, $b_i a_i > 0$.

Suppose first that the projection of the edge a_0a is free of double points, so that no point of V lies directly above (or below) a_0a. By choosing $b \in I(a_0, a)$ to be sufficiently far above (or below) E^2 we can make the angle between the plane a_0ba and the plane E^2 arbitrarily close to $\pi/2$. It then follows that we can always find a point $b \in I(a_0, a)$ such that $[a_0, b, a]V = [a_0, a]$, hence $\mathcal{S}^b_{a_0a}$ will be applicable to V. Moreover, $\mathcal{S}^{b_1}_{a_0a}$ will also be applicable to V for all points b_1 which lie directly above (respectively below) b. Since the simplex S satisfies hypotheses (i), (ii) and (iii), each such triangle $a_0b_1a_1$ will meet S in at most a (possibly degenerate) triangle, and by moving b_1 still further above or below b we can arrange it so that the sawtooth avoids S.

In the general case, the edge a_0a may contain a finite set of points, say $p_1, \cdots, p_r$, which project to double points in the link diagram. If this occurs, it may not be possible to choose a single point b_1 such that $\mathcal{S}^{b_1}_{a_0a}$ is applicable to V, and our construction will be more complicated. Our idea will be to first place sharp teeth about each double point, making each such tooth sufficiently narrow so that it can be slipped in between the edges of the link, and to then complete the construction by the method used before. We consider the points $p_1, \cdots, p_r$ one at a time. Let P_1 be a plane through p_1 and ℓ. This plane meets S in a convex set, hence it is always possible to draw a line from p_1 to ℓ, in the plane P_1, which avoids both the link and S. Thickening the line, we may produce a triangle with base $[a_1, a_2] \subset [a_0, a]$ and vertex $b_2 \in I(a_1, a_2)$, such that the triangle $[a_1, b_2, a_2]$ avoids S and meets V in $[a_1, a_2]$. Observe that the segment $[a_0, a_1]$ is free of double points, hence the earlier construction may now be applied to locate a point $b_1 \in I(a_0, a_1)$ such that $[a_0, b_1, a_1] \cup S = \emptyset$ or a_0, and $[a_0, b_1, a_1]V = [a_0, a_1]$. Repeating the construction on the points $p_2, \cdots, p_r$ we may locate points a_{2i-1}, a_{2i}, b_{2i-1}, b_{2i} for each $i = 2, \cdots, r$ and finally a single point b_{2r+1}.

Then $\mathcal{S}^{b_1 \cdots b_{2r+1}}_{a_0 \cdots a_{2r+1}}$ will be applicable to V and will avoid S. This completes the proof of Lemma 2.1.2. ||

The connection between braids and links which is given in Theorem 2.1 can be exploited to characterize the fundamental groups of the complements of all possible links in S^3:

THEOREM 2.2 [Artin, 1925; the proof given here is different from that given by Artin]. *Let $\beta \in B_n$, and suppose that the action of β on the free group F_n is given by equations (1-22). Let $\hat{\beta}$ be the link determined by the braid β. Then the fundamental group $\pi_1(S^3 - \hat{\beta})$ of the complement of $\hat{\beta}$ in S^3 admits the presentation:*

(2-3)
generators: $y_1, \cdots, y_n$

defining relations: $y_i = A_i(y_1, \cdots, y_n) y_{\mu_i} A_i^{-1}(y_1, \cdots, y_n) \quad 1 \le i \le n-1.$

Moreover, every link group admits such a presentation.

Proof. The reader is referred to Theorem 1.10 of Chapter 1. The disc $D^2 - Q_n$ will be assumed to have the meaning given it there (cf. Figure 5), and the braid group B_n will be interpreted as the group of automorphisms of $\pi_1(D^2 - Q_n)$ which are induced by topological mappings which keep ∂D^2 fixed pointwise.

Let $b : (D^2 - Q_n) \to (D^2 - Q_n)$ be a topological mapping which induces the automorphism $\beta \in B_n \subset \mathrm{Aut}\, \pi_1(D^2 - Q_n)$. The action of β on the generators $x_1, \cdots, x_n$ of $\pi_1(D^2 - Q_n)$ will be assumed to be given by equations (1-22). Then b may be used to construct a manifold $(D^2 - Q_n) \times I/\sim$, defined as the quotient space of $(D^2 - Q_n) \times I$ obtained by identifying points on $(D^2 - Q_n) \times 0$ with points on $(D^2 - Q_n) \times 1$ by the rule $(z, 1) = (b(z), 0)$. Our manifold may be visualized geometrically as the complement of a link in the solid torus $T = D^2 \times I/\sim$. We denote the quotient space $(D^2 - Q_n) \times I/\sim$ by $(T - \hat{\beta})$. In view of Theorem 2.1, the complement of every link in a solid torus may be obtained in this way, by allowing β to range over B_n and n to range over the positive integers.

Our manifold $(T-\hat{\beta})$ clearly fibers over the circle S^1, the fiber being (D^2-Q_n). The exact sequence of the fibration[4] then gives a short exact sequence:

$$1 \to \pi_1(D^2-Q_n) \to \pi_1(T-\hat{\beta}) \to \pi_1 S^1 \to 1 \tag{2-4}$$

where the trivial group on the left is $\pi_2(S^1)=1$, and the trivial group on the right is $\pi_0(D^2-Q_n)=1$.

Let $y_1,\cdots,y_n$ denote the images of $x_1,\cdots,x_n$ under the natural embedding of $\pi_1(D^2-Q_n)=\pi_1((D^2-Q_n)\times\{0\})$ in $\pi_1(T-\hat{\beta})$. Let $t\in\pi_1(T-\hat{\beta})$ be represented by $z\times I$, $z\in\partial D^2$, oriented from $\partial D^2\times\{1\}$ to $\partial D^2\times\{0\}$. (Thus t is a longitude on the boundary of the solid torus T.) Note that t is also a lift of a generator of $\pi_1 S^1$ to $\pi_1(T-\hat{\beta})$. Then the short exact sequence above defines a presentation for $\pi_1(T-\hat{\beta})$, with generators $y_1,\cdots,y_n,t$ and defining relations:

$$ty_it^{-1} = A_i(y_1,\cdots,y_n)y_{\mu_i}A_i^{-1}(y_1,\cdots,y_n), \quad 1\le \ell\le n\,. \tag{2-5}$$

To obtain from this a presentation for the complement of the same link $\hat{\beta}$ in S^3, let m be a meridian on the boundary of $(T-\hat{\beta})$, and let M and L be a meridian and longitude on the boundary of (S^3-T). Then, by Van Kampen's theorem, the fundamental group of $(S^3-\hat{\beta})=(T-\hat{\beta})\cup(S^3-T)$, where $(T-\hat{\beta})\cap(S^3-T)=\partial T$, admits the presentation:

(2-6) *generators* $y_1,\cdots,y_n,m,L,M,t$

defining relations $ty_it^{-1} = A_i(y_1,\cdots,y_n)y_{\mu_i}A_i^{-1}(y_1,\cdots,y_n)\ 1\le i\le n$

$$M=1$$
$$t=M$$
$$m=L$$
$$m=y_1y_2\cdots y_n\,.$$

4 See, for example, Hilton, "An introduction to homotopy theory," Cambridge University Press 1963, p. 55.

A simple sequence of Tietze transformations now shows that the presentation defined by (2-6) is equivalent to the presentation defined by (2-3). The relation $y_n = A_n(y_1,\cdots,y_n)y_{\mu_n}A_n^{-1}(y_1,\cdots,y_n)$ has been omitted because condition (1-24) implies that it is a consequence of the remaining relations. ||

COROLLARY 2.2.1. *Let* F_n *be a free group of rank* n, *freely generated by* $x_1,\cdots,x_n$. *Let* $(\mu_1,\cdots,\mu_n)$ *be a permutation of the letters* $1,\cdots,n$, *and let* $A_1,\cdots,A_n$ *(where* A_i *denotes* $A_i(x_1,\cdots,x_n)$*) be any set of words in the generators* $x_1,\cdots,x_n$ *of* F_n *which satisfy the free equality (1-24). Then the abstract group defined by the presentation (2-2) is the fundamental group of the complement of a link in* S^3. *Conversely, every link group admits a presentation of this type.*

Proof. By Theorem 1.9, the endomorphism β from F_n to F_n defined by $(x_i)\beta = A_i x_{\mu_i} A_i^{-1}$, $1 \le i \le n$, is a braid automorphism of F_n. By Theorem 1.8 and Corollary 1.8.3 the braid automorphism β can be realized by a geometric braid. By Theorem 2.2 the fundamental group $\pi_1(S^3 - \hat{\beta})$ of the complement of the closed braid $\hat{\beta}$ in S^3 admits the presentation (2-2).

Conversely, if V is an arbitrary link, then by Theorem 2.1 there exists a closed braid $\hat{\beta}$ which represents V. This closed braid can be associated (non-uniquely) with an open braid β. The open braid β can be associated with a unique braid automorphism, and this braid automorphism determines the presentation (2-2) for $\pi_1(S^3 - \hat{\beta})$. ||

2.2. *Markov's Theorem*

The theorem which will be stated and proved in this section (Theorem 2.3) will allow us to translate the question of whether two links are equivalent into a purely algebraic question (Corollary 2.3.1) about the collection of braid groups $\{B_1, B_2, B_3, \cdots\}$. This theorem was announced by Markov in 1935 in a report which gave the broad outlines of a proof, but omitted details. The details were never published. Several years later, there was another brief announcement [Weinberg, 1939] of an improved version of Markov's Theorem. The heart of Markov's asserted proof rests on a long and exacting calculation, requiring the patient enumeration of a large number of separate cases. The proof we give here is based on notes taken at a seminar at Princeton University in 1954. The speaker is unknown to us; we thank him for his help!

The basic idea behind Markov's theorem is to replace the notion of combinatorial equivalence of closed braids by a more restrictive notion, braid equivalence, which guarantees that the intermediate links in a sequence of deformations will always be closed braids. This is accomplished by replacing the type $\mathcal{E}$ deformations introduced in Section 2.1, which do not necessarily preserve closed braids, by alternative types of deformations (each of which is an appropriate product of type $\mathcal{E}$ deformations), which do preserve closed braids. Before stating the theorem, it will be necessary to define these operations and to establish certain relationships between them. The type $\mathcal{S}$ operations, also introduced in Section 2.1, will play an important role in the proof of Theorem 2.3, but do not appear in the statement.

The symbols $V, V', V^*, V^{\#}, \hat{V}, V_i, \cdots$ will be used to denote an oriented link which is in general position with respect to a braid axis ℓ. This link is not, in general, a closed braid. In our link diagrams, the symbol ↻ will be used to denote the intersection of the axis with the plane of projection, with the arrow indicating the positive sense of rotation about ℓ. In certain of our link diagrams we will wish to indicate the edges of a link both before and after a deformation, and for this purpose we will use dashed lines to indicate the original edges and solid lines to indicate their replacements.

DEFINITION. If $\mathcal{E}^b_{ac}$ is applicable to V, and if $ab > 0$, $bc > 0$, $ac > 0$ then we shall write

$$\mathcal{R}^b_{ac}V = \mathcal{E}^b_{ac}V = V - ac + ab + bc. \qquad (2\text{-}7)$$

This operation and its inverse will be called type $\mathcal{R}$ (see Figure 8).

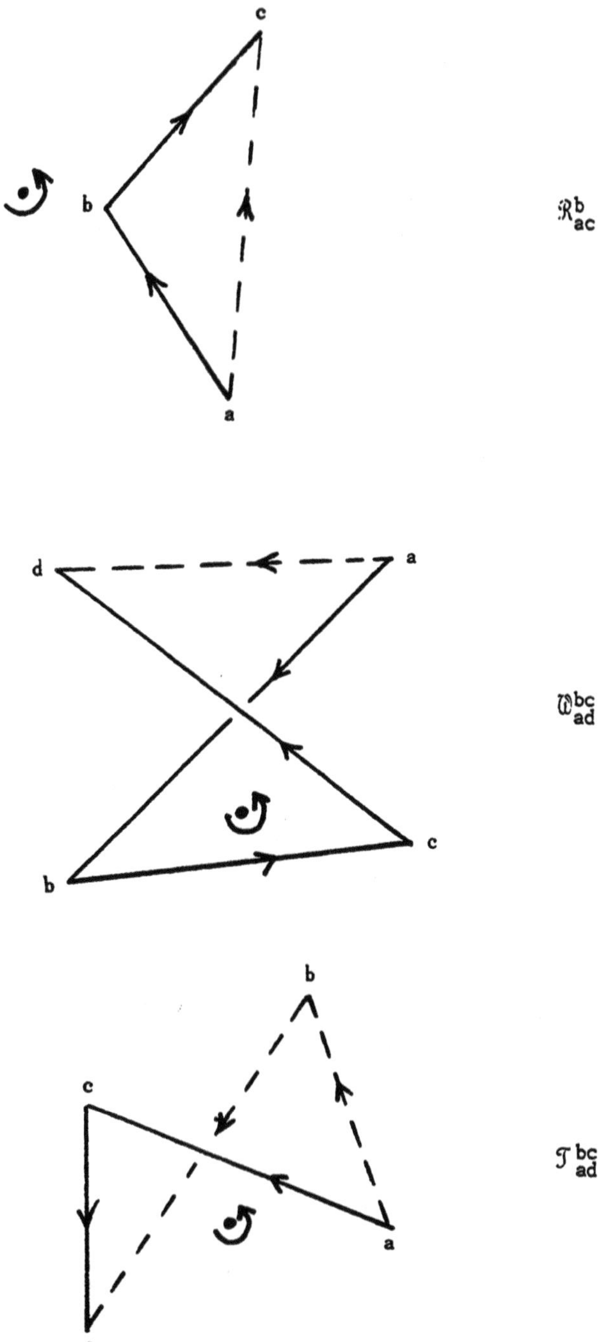

Fig. 8.

Suppose that $\mathfrak{E}^{b}_{ad}$ is applicable to V and that $\mathfrak{E}^{c}_{bd}$ is applicable to $\mathfrak{E}^{b}_{ad}V$. If $ab > 0$, $bc > 0$, $cd > 0$, $ca > 0$, $db > 0$, $ad > 0$ and if $[a, b, c, d][V] = [a, d]$ then we set

$$\mathfrak{W}^{bc}_{ad}V = \mathfrak{E}^{c}_{bd}\,\mathfrak{E}^{b}_{ad}V = V - ad + ab + bc + cd \,. \tag{2-8}$$

We call $\mathfrak{W}^{bc}_{ad}$ and its inverse operations of *type* $\mathfrak{W}$. (See Figure 8.)

We may now state Markov's Theorem.

THEOREM 2.3 [Markov, 1935]. *Let* V, V′ *be two closed braids which are combinatorially equivalent.*[5] *Then there exists a finite sequence of closed braids* $V = V_0, V_1, V_2, \cdots, V_s = V'$ *such that each* V_i, $1 \leq i \leq s$, *is obtained from* V_{i-1} *by a single application of an operation of type* $\mathfrak{R}$ *or of type* $\mathfrak{W}$.

Theorem 2.3 may be interpreted as a strong version of Theorem 2.1. Recall that Theorem 2.1 says that every link type may be represented (non-uniquely) by a closed braid. Theorem 2.3 describes how the various closed braid representatives of a given link type are related to one another. We will show later (Corollary 2.3.1) that if we choose an open braid representative of a given closed braid, say $\beta_i \in B_n$ representing V_i, then the effect on β_i of a type $\mathfrak{R}$ operation acting on V_i may be realized algebraically by conjugation within the braid group, while the effect of type $\mathfrak{W}$ operations may be realized by altering β_i and string index n in a particularly simple way (roughly speaking).

The proof of Theorem 2.3 will occupy most of the remainder of this section. While simple in concept, it involves a large number of details, some of a computational nature, with certain of these being rather delicate. Lemmas 2.3.1-2.3.3 are technical in nature, and will serve to

---

5 Note that V and V′ will not, in general, have the same string index.

reduce the computational parts of the proof to manageable proportions. Lemmas 2.3.4 and 2.3.5 are the heart of the matter. Following the statement of Lemmas 2.3.4 and 2.3.5 we will establish that the truths of Lemmas 2.3.4 and 2.3.5 implies the truth of Theorem 2.3. We will then prove Lemmas 2.3.4 and 2.3.5.

DEFINITION. Let V and V′ be links. A sequence of links $V = V_0 \to V_1 \to \cdots \to V_n = V'$ which have the property that each $V_i = \mathfrak{F}_i V_{i-1}$, where $\mathfrak{F}_i$ is a single operation of type $\mathfrak{E}, \mathfrak{R}, \mathfrak{W}$ or $\mathfrak{S}$, will be called a *deformation chain* joining V to V′. Observe that if V and V′ are combinatorially equivalent (cf. Section 2.1) then it is trivially true that there is always a deformation chain joining V to V′, since V may be deformed to V′ by a finite sequence of type $\mathfrak{E}$ operations.

LEMMA 2.3.1. *If* V^* *and* V^{**} *are in general position and are combinatorially equivalent, then there is deformation chain from* V^* *to* V^{**} *such that all links in the chain are in general position.*

Proof of Lemma 2.3.1. If $\mathfrak{E}^b_{ac}$ creates an edge which is coplanar with ℓ we replace it by $\mathfrak{E}^{b'}_{ac}$ where b′ is close to b but not in the plane of [a, b, c]. Afterward replace b by b′ in all links which follow. If $(\mathfrak{E}^b_{ac})^{-1}$ creates an edge ac which is coplanar with ℓ, and if cd is the edge following bc, choose a point c′ close to c and on bc, replace $(\mathfrak{E}^b_{ac})^{-1}$ by $(\mathfrak{E}^c_{c'd})^{-1}(\mathfrak{E}^b_{ac'})^{-1}(\mathfrak{E}^{c'}_{bc})$, and replace c by c′ in all links of the deformation chain. This will create a new deformation chain joining V to V′ which has two more links than the old one, but one less edge coplanar with ℓ. Induction on the number of edges coplanar with ℓ completes the proof. ||

In view of Lemma 2.3.1, we may assume without loss of generality that all links in a deformation chain are in general position. This assumption will not be repeated explicitly in the lemmas which follow.

To allow greater flexibility in the proof, we now introduce one more special type of link deformation, denoted type $\mathfrak{T}$. We will then show (Lemma 2.3.2 below) that an operation of type $\mathfrak{T}$ may always be expressed in terms of operations of type $\mathfrak{R}$ and $\mathfrak{W}$. Thus any time that we use an operation of type $\mathfrak{T}$, it will be equivalent to a sequence of operations of type $\mathfrak{R}$ and $\mathfrak{W}$.

Suppose that $(\mathfrak{E}^{b}_{ad})^{-1}$ is applicable to V, and that $\mathfrak{E}^{c}_{ad}$ is applicable to $(\mathfrak{E}^{b}_{ad})^{-1}V$. If $ab > 0$, $ac > 0$, $bd > 0$, $cd > 0$, $ad < 0$; then we define

$$\mathfrak{T}^{bc}_{ad}V = \mathfrak{E}^{c}_{ad}(\mathfrak{E}^{b}_{ad})^{-1}V = V - ab - bd + ac + cd \, . \tag{2-9}$$

We call $\mathfrak{T}^{bc}_{ad}$ and its inverse operations of *type* $\mathfrak{T}$. (See Figure 8.)

LEMMA 2.3.2 [Weinberg, 1939]. *Any operation of type* $\mathfrak{T}$ *may be expressed in terms of operations of types* $\mathfrak{R}$ *and* $\mathfrak{W}$.

Proof. To see that a type $\mathfrak{T}$ operation (which has been defined in terms of type $\mathfrak{E}$ operations) can be expressed as a product of operations of type $\mathfrak{R}$ and $\mathfrak{W}$, we show that there exist points a', d' such that

$$\mathfrak{T}^{bc}_{ad}V = (\mathfrak{R}^{d'}_{cd})^{-1}(\mathfrak{W}^{a'b}_{d'd})^{-1}\mathfrak{W}^{cd'}_{aa'}\mathfrak{R}^{a'}_{ab}V \tag{2-10}$$

is applicable. We may choose the point a' in the plane of $[a, b, d]$ close to a and with $aa' > 0$. Since $ab > 0$, we have $a'b > 0$, thus $\mathfrak{R}^{a'}_{ab}$ is applicable. Choose d' in the plane of $[a, c, d]$ close to d. Since $([a, b, d] + [a, c, d])V = [a, b] + [b, d]$ and since a' and d' are close to a and d respectively, we have

$$[a, c, d', a'][\mathfrak{R}^{a'}_{ab}V] = [a, a'] \, .$$

Also $ac > 0$, $cd' > 0$, $d'a' > 0$, $d'a > 0$, $a'c > 0$, and $aa' > 0$; hence $\mathfrak{W}^{cd'}_{aa'}$ is applicable to $\mathfrak{R}^{a'}_{ab}V$. Similarly, $(\mathfrak{W}^{a'b}_{d'd})^{-1}$ and $(\mathfrak{R}^{d'}_{cd})^{-1}$ are applicable. ||

We now have at our disposal 5 types of link deformations:

Type $\mathfrak{E}$: The basic elementary combinatorial deformation. A type $\mathfrak{E}$ operation may (or may not) alter height $h(V)$.

Type $\mathfrak{R}$: A special type $\mathfrak{E}$ deformation which always leaves $h(V)$ unaltered.

Type $\mathfrak{W}$: Leaves $h(V)$ unaltered.

Type $\mathfrak{T}$: Leaves $h(V)$ unaltered.

Type $\mathfrak{S}$: Always alters $h(V)$ by precisely 1.

The lemma which follows develops certain properties of the operation $\mathfrak{S}^{b_1 \cdots b_m}_{a_0 \cdots a_m}$ which will be useful later.

LEMMA 2.3.3. *The operation* $\left(\mathfrak{S}^{d_1 \cdots d_m}_{c_0 \cdots c_m}\right)\left(\mathfrak{S}^{b_1 \cdots b_n}_{a_0 \cdots a_n}\right)^{-1}$, *where* $[a_0, a_n] = [c_0, c_m]$, *may be expressed in terms of operations of type* $\mathfrak{R}$ *and* $\mathfrak{W}$.

Proof. First we show that if we have a sawtooth on a_0a_n we may always construct a finer sawtooth on a_0a_n having a given point $d \in [a_0, a_n]$ lying on the sawtooth, and using only operations of types $\mathfrak{W}$ and $\mathfrak{T}$. Let $d \in [a_{i-1}, a_i]$. If $b_id > 0$, pick a point e such that $de > 0$, $ea_i > 0$, with e close enough to $[a_{i-1}, b_i, a_i]$ so that $\mathfrak{W}^{de}_{b_i, a_i}$ is applicable to V (see Figure 9). This operation produces the finer sawtooth. If $b_id < 0$, we pick e such that $a_{i-1}e > 0$, $ed > 0$, and apply $\mathfrak{W}^{ed}_{a_{i-1}, b_i}$. Finally, if $b_id = 0$, we move b_i by an operation of type $\mathfrak{T}$. Hence if we have a sawtooth at a_0a_n we may produce a finer sawtooth at any finite set of points by operations of type $\mathfrak{W}$ and $\mathfrak{T}$. Clearly the process is reversible.

To perform $\left(\mathfrak{S}^{d_1 \cdots d_m}_{c_0 \cdots c_m}\right)\left(\mathfrak{S}^{b_1 \cdots b_n}_{a_0 \cdots a_n}\right)^{-1}$ we construct a finer sawtooth at $[a_0, a_n]$ intersecting $[a_0, a_n]$ at the points $\{a_0, \cdots, a_n\} \cup \{c_0, \cdots, c_m\}$, use operations of type $\mathfrak{T}$ to place each tooth properly and reverse the process of producing a fine sawtooth. Since, by Lemma 2.3.2, an operation of type $\mathfrak{T}$ may be replaced by a sequence of operations of type $\mathfrak{R}$ and $\mathfrak{W}$, the statement of Lemma 2.3.3 follows. ||

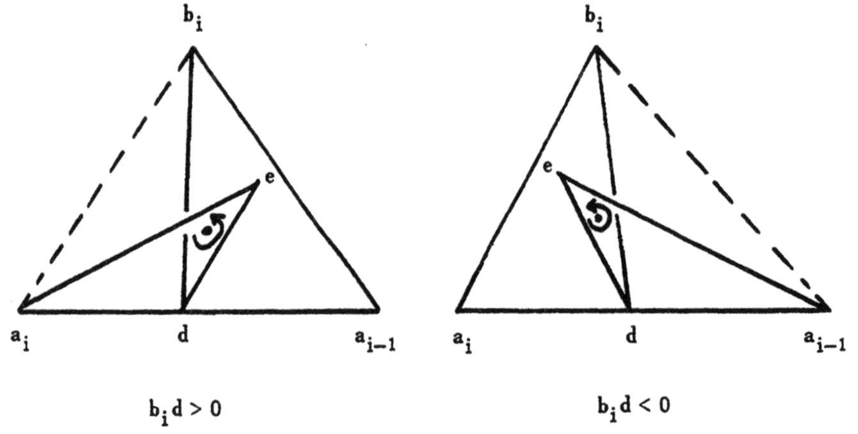

Fig. 9.

The preparatory material is now complete, and we are now ready to state and prove the two lemmas which together imply the truth of Theorem 2.3. Recall that if V is a link in general position, then the height of V, denoted $h(V)$, is the number of negative edges of V. Thus $h(V) = 0$ if and only if V is a closed braid.

LEMMA 2.3.4. *If* $V^* \to V^{**}$ *is a deformation chain, and if* $h(V^*) = h(V^{**}) > 0$, *then there exists a second deformation chain* $V^* \to V_1 \to \cdots \to V_s \to V^{**}$ *such that* $s \geq 1$ *and also* $h(V_i) < h(V^*)$ *for every* $i = 1, \cdots, s$.

LEMMA 2.3.5. *If* $V^* \to \hat{V} \to V^{**}$ *is a deformation chain, and if* $h(V^*) < h(\hat{V})$, *and also* $h(V^{**}) < h(\hat{V})$, *then there exists a second deformation chain* $V^* \to V_1 \to \cdots \to V_s \to V^{**}$ *such that* $s \geq 1$ *and also* $h(V_i) < h(\hat{V})$ *for every* $i = 1, \cdots, s$.

Lemma 2.3.5 says that if two points in a "valley" may be joined by a path that goes over the "mountain," and if at some point a subpath contains a local peak, then this subpath may always be replaced by another (possibly longer) subpath with a lower peak. Lemma 2.3.4 says that if a subpath is level, then this subpath may be replaced by a new (longer) path which dips down, thereby creating two potential new peaks. Together these lemmas will imply that if any two points in the valley can be joined by a path which goes over the mountain, then there is another (longer) path joining them which remains always in the valley.

Before proving Lemmas 2.3.4 and 2.3.5 we show that they imply the truth of Theorem 2.3. By hypothesis, V and V' are closed braids which are combinatorially equivalent. Hence there is a deformation chain $V \to V_1 \to V_2 \to \cdots \to V_r \to V'$ joining V to V'. By Lemma 2.3.1, each V_i in this chain may be assumed to be in general position. Hence the height $h(V_i)$ of each link in the chain is well-defined.

It may happen that $h(V_i) = 0$ for every $i = 1, \cdots, r$. In this case every link in the chain is a closed braid. It then follows that the deformations which join the link V_{i-1} to $V_i (i = 2, \cdots, r)$ must have all been type $\mathfrak{R}$ and type $\mathfrak{W}$, since operations of type $\mathfrak{E} \neq \mathfrak{R}$ or of type $\mathfrak{S}$ either create negative edges or require the existence of a negative edge, and since none of the links in the chain has a negative edge, it follows that the chain only admits operations of type $\mathfrak{R}$ or $\mathfrak{W}$.

If $h(V_i) \neq 0$ for some $i = 1, \cdots, r$ let $h = \max h(V_i) > 0$. Now, there may be some j for which $h(V_{j-1}) = h(V_j) = h$. By Lemma 2.3.4, the chain can then be replaced by a new chain $V \to V_1^{\#} \to \cdots \to V_e^{\#} \to V'$ joining V to V', where the new chain has the property that $h = \max h(V_i^{\#}) > 0$, but no j exists with $h(V_{j-1}^{\#}) = h(V_j^{\#}) = h$.

Now we must be in the situation of Lemma 2.3.5. Using Lemma 2.3.5, we next construct another deformation chain $V \to \bar{V}_1 \to \cdots \to \bar{V}_q \to V'$ which likewise connects V to V', but such that $\max h(\bar{V}_i) < h (i = 1, \cdots, q)$. Continuing in this way, we ultimately obtain a deformation chain joining V to V' such that the height of every link in the chain is zero. In this final chain, only operations of type $\mathfrak{R}$ and $\mathfrak{W}$ are admissible. This completes the proof of Theorem 2.3, assuming the truth of Lemmas 2.3.4 and 2.3.5. It remains to establish Lemmas 2.3.4 and 2.3.5.

Proof of Lemma 2.3.4. By hypothesis, $V^{**} = \mathfrak{F}V^*$, where $\mathfrak{F}$ is a single operation of type $\mathfrak{E}, \mathfrak{R}, \mathfrak{W}$ or $\mathfrak{S}$. Operations of type $\mathfrak{S}$ necessarily alter $h(V^*)$, hence $\mathfrak{F}$ cannot be type $\mathfrak{S}$. Operations $\mathfrak{R}$ and $\mathfrak{W}$ leave $h(V^*)$ unaltered and are therefore admissible. Operations of type $\mathfrak{E}$ may or may not change $h(V^*)$, and it is necessary to consider the various cases separately. We distinguish between the 8 different possibilities for type $\mathfrak{E}$, denoted $\mathfrak{E}^k$ or $\mathfrak{E}^{bk}_{ac}$, where $k = 0, \cdots, 7$, according to the table below:

Table 1

k	ac	ab	bc	$h(\mathfrak{E}^b_{ac}V^*) - h(V^*)$	$h((\mathfrak{E}^b_{ac})^{-1}V^*) - h(V^*)$
0	+	+	+	0	0
1	–	+	+	–1	+1
2	+	+	–	+1	–1
3	–	+	–	0	0
4	+	–	+	+1	–1
5	–	–	+	0	0
6	+	–	–	+2	–2
7	–	–	–	+1	–1

Since $\mathfrak{E}^0$ is type $\mathfrak{R}$ and $\mathfrak{E}^1$ type $\mathfrak{S}$, we may restrict our attention to $k = 2, \cdots, 7$. Of these, only $\mathfrak{E}^3$ and $\mathfrak{E}^5$ leave $h(V^*)$ unaltered, hence $\mathfrak{F}$ must be either type $\mathfrak{R}, \mathfrak{W}, \mathfrak{E}^3$ or $\mathfrak{E}^5$, or the inverse of any of these.

If $\mathfrak{F}$ is type $\mathfrak{R}^{\pm 1}$ or type $\mathfrak{W}^{\pm 1}$, choose any negative edge a_0a of V^* (one always exists because $h(V^*) > 0$) and construct a sawtooth on a_0a by the method of Lemma 2.1.2, avoiding the triangle formed by $\mathfrak{R}$ or the tetrahedron of $\mathfrak{W}$. The sawtooth may be used to define an operation of type $\mathfrak{S}$, and we may replace $\mathfrak{F}$ by $\mathfrak{S}^{-1}\mathfrak{F}\mathfrak{S}$. The first deformation $\mathfrak{S}$ reduces the height, the second leaves it unaltered, and the third increases it, thus satisfying the conditions of the lemma.

If $\mathfrak{F}$ is type $(\mathfrak{E}^k)^{\pm 1}$, $k = 3$ or 5, and if $h(V^*) > 1$, a similar argument applies, because each of these type $\mathfrak{E}$ operations involves a single negative edge V^*, hence if $h(V^*) > 1$ we may choose any other negative edge of V^*, and use it to define an operation $\mathfrak{S}$ (avoiding the triangle formed by $\mathfrak{E}$). We then replace $\mathfrak{F}$ by $\mathfrak{S}^{-1}\mathfrak{F}\mathfrak{S}$. Hence the lemma is easily established whenever $h(V^*) > 1$. If $h(V^*) = 1$ we must work a little harder to take care of the cases $\mathfrak{F} = (\mathfrak{E}^k)^{\pm 1}$, $k = 3$ or 5.

If $\mathfrak{F} = \mathfrak{E}^{b3}_{ac}$, we first consider the case where we can erect sawteeth on ac and bc which share points $(b_1, \cdots, b_m)$ (see Figure 10). This will always be possible if ac and bc are very close together. We shall denote the subdivisions on ac by a_i and those on bc by c_i, with $a_0 = a$, $c_0 = b$. These sawteeth shall be such that the pyramids

$$[a_i, a_{i+1}, c_i, c_{i+1}, b_{i+1}]\, V^* = [a_i, a_{i+1}] \, .$$

This is always possible because we have assumed that ac is close to bc. Also, we require that $a_i c_i \neq 0$, that is the edge $a_i c_i$ does not lie on a line which is coplanar with ℓ. A calculation shows that $\mathfrak{F}$ can then be replaced by the sequence of deformations:

$$\mathfrak{F} = \left(\mathfrak{S}^{b_1 \cdots b_m}_{c_0 \cdots c_m}\right)^{-1} \mathfrak{F}_0 \mathfrak{F}_1 \cdots \mathfrak{F}_{m-1} \mathfrak{S}^{b_1 \cdots b_m}_{a_0 \cdots a_m} \tag{2-11}$$

where

$$\mathfrak{F}_i = \left(\mathfrak{R}^{a_i}_{b_i c_i}\right)^{-1} \left(\mathfrak{R}^{c_i}_{a_i b_{i+1}}\right) \quad \text{if } a_i c_i > 0 \text{ and } i \neq 0 \, ,$$

$$= \left(\mathfrak{R}^{a_i}_{c_i, b_{i+1}}\right)^{-1} \left(\mathfrak{R}^{c_i}_{b_i a_i}\right) \quad \text{if } a_i c_i < 0 \text{ and } i \neq 0$$

$$\mathfrak{F}_0 = \mathfrak{R}^{c_0}_{a_0 b_1} = \mathfrak{R}^{b}_{ab_1} \, .$$

In the replacement sequence the first $\mathfrak{S}$ operation replaces the negative edge ac by a sawtooth on ac, thereby reducing the height. The operation

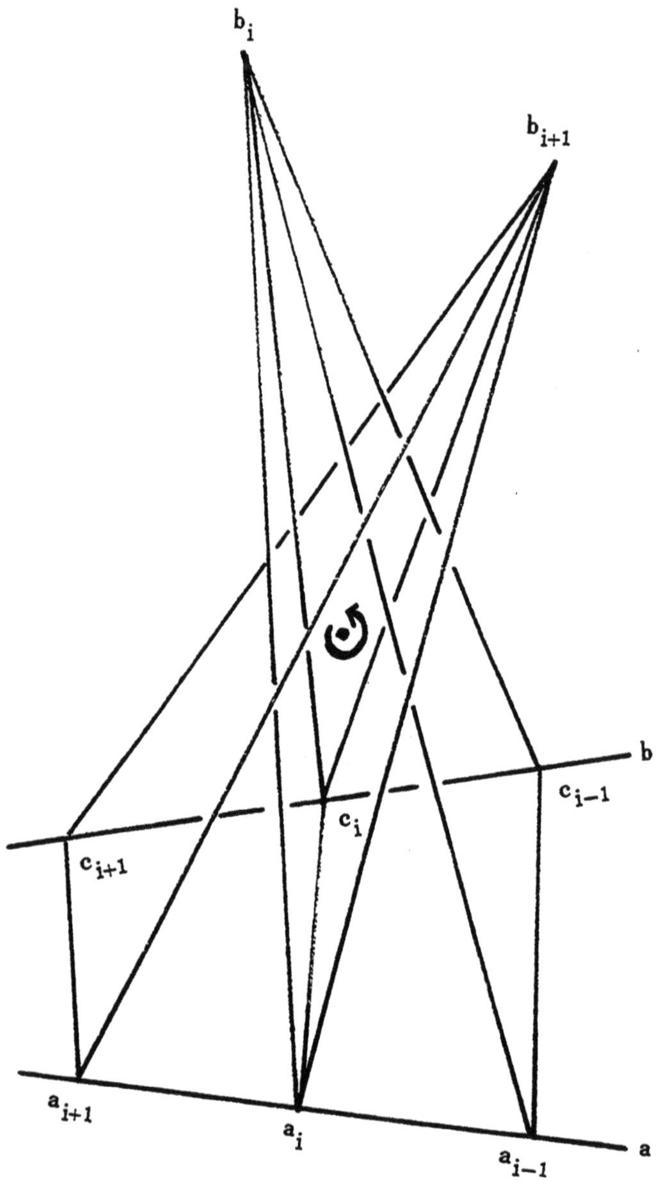

Fig. 10

operation $\mathcal{F}_i$ may be visualized as sliding a single vertex a_i of the link across the triangle abc to a new position at c_i. The product $\mathcal{F}_0\mathcal{F}_1\cdots\mathcal{F}_{m-1}$ thus replaces the sawtooth on ac by a sawtooth on bc (plus the positive edge ab). Each $\mathcal{F}_i$ operation leaves height invariant. The final $\mathcal{S}^{-1}$ operation replaces the sawtooth on bc by a negative edge on bc, thereby increasing the height again. Thus the requirements of the lemma are satisfied.

In the general case, we will show that there exist a finite set of points $a_{00}, a_{01}, \cdots, a_{0r} \in [a, b]$, with $a_{00} = a$, $a_{0r} = b$, such that every adjacent pair of lines $a_{0,j}c$ and $a_{0,j-1}c$, $j = 1, \cdots, r$, admit a common sawtooth (just as ac and bc did in the special case considered above). It will then be possible to replace $\mathcal{F}$ by a product of the form

$$\mathcal{F} = \prod_{j=1}^{r} (\mathcal{S}'_j)^{-1}\mathcal{F}_{0j}\mathcal{F}_{1j}\cdots\mathcal{F}_{m-1,j}\mathcal{S}_j$$

where each $\mathcal{S}_j, \mathcal{S}'_j$ is an appropriate type $\mathcal{S}$ operation and each $\mathcal{F}_{ij}$ is a product of two type $\mathcal{R}$ operations. Each $\mathcal{S}_j(\mathcal{S}'_{j-1})^{-1}$ may by Lemma 2.3.3 be replaced by operations of type $\mathcal{R}$ and $\mathcal{W}$. Hence $\mathcal{F} = (\mathcal{S}'_r)^{-1}\mathcal{M}\mathcal{S}_1$ where $\mathcal{M}$ is a product of operations of type $\mathcal{R}$ and $\mathcal{W}$ as required.

In order to show that the required sequence of points $\{a_{0j}\}$ exist, we examine the projection of the link V^* onto the plane E^2 (cf. Section 2.1, proof of Lemma 2.1.2). We now study carefully the constraints which exist when we attempt to construct a sawtooth on a negative edge of V^*.

1. Let rs be a negative edge of V^*. In general, rs will contain a finite set of double points, and we restrict ourselves initially to the situation where all double points correspond to undercrossings. Then, we assert that it is always possible to find a point q such that $rq > 0$, $qs > 0$, and also $[r, q, s]V^* = [r, s]$. To see why this is so, observe that $rq > 0$, $qs > 0$ will be satisfied for any point q which lies in the wedge of 3-space which is labeled I(rs) in Figure 7 (here $r = a_{i-1}$, $s = a_i$). Moreover, the angle between the plane of projection and the triangle $[r, s, q]$ can be made arbitrarily close to $\pi/2$ by choosing q to be

sufficiently far above the plane of projection. Since no edge of V^* lies directly above rs, it is thus clear that by choosing q to be sufficiently far above the plane of projection we can ensure that $[r, q, s]\, V^* = [r, s]$.

2. More generally, let rs and $r's'$ be a pair of coplanar negative edges of V^* which have the property that no other edges of the link lie above the quadrilateral $[r, s, s', r']$. Also, it will be assumed that the axis ℓ does not pierce $[r, s, s', r']$. Then we assert that it is always possible to find a point q such that

(i) $rq > 0, \; qs > 0$

(ii) $r'q > 0, \; qs' > 0$

(iii) $[r, s, s', r', q]\, V^* = [r, s] \cup [r', s']$.

The construction is identical to the previous case. Each edge rs, or $r's'$, determines a region $I(rs)$, $I(r', s')$ in 3-space from which admissible points q may be chosen to satisfy conditions (i) and (ii) respectively. The intersection of these regions is necessarily non-empty, because ℓ does not pierce $[r, s, s', r']$. To satisfy (iii) we choose q sufficiently far above the plane of projection. Since no edges of V^* lie above $[r, s, s', r']$ we can always find a suitable point q.

In the discussion above, if all edges of V^* lie above $[r, s, s', r']$ we can again find q sufficiently far below the plane of projection.

3. Let rs and $r's'$ be a pair of coplanar negative edges, which have the property that $[r, s, s', r']\, V^* = [r, s] \cup [r', s']$, that $[r, s, s', r']$ does not contain ℓ, and also that V^* contains precisely one edge (say $\mathfrak{k}$) which lies entirely above $[r, s, s', r']$ and precisely one edge (say $\mathfrak{k}'$) which lies entirely below $[r, s, s', r']$ so that the projection of $[r, s, s', r']$ contains one double point, where $\mathfrak{k}$ crosses over $\mathfrak{k}'$. Let $p \in [r, s, s', r']$ correspond to this double point. It will be assumed that r, s, s', r' are all very close to p. Then we claim that as before we can always find a point q which satisfies condition (i), (ii) and (iii) above. To locate q, construct the plane through the axis ℓ and the point p. The link is in general position, hence it meets this plane in a finite set of points. Hence we can always draw a line from p to ℓ which avoids the link, and thicken

it to produce a cone with vertex q which avoids the link. The vertex of the cone will be in the region $I(r, s) \cap I(r', s')$, as before. The base of this cone will cover the quadrilateral $[r, s, s', r']$ if the points r, s, s', r' are sufficiently close to p. Hence, by construction, conditions (i), (ii) and (iii) will be satisfied. (This situation differs from that in (2) above, because instead of choosing q far above or below the plane of projection, we squeeze the pyramid $[r, s, s', r', q]$ into the available space between the other link edges.)

4. We now consider the general case. Let ac be a negative edge of the link V^*, and suppose that $\mathcal{E}_{ac}^{b3}$ is applicable. Then $[a, b, c]V^* = ac$. Since there are only a finite number of double points in the triangle $[a, b, c]$, it will always be possible to subdivide $[a, b, c]$ into a network of subtriangles $[a_{0,j-1}, a_{0,j}, c]$, where $a_{00} = a$, $a_{0r} = b$, and each $a_{0j} \in [a, b]$, and to subdivide each such triangle into a network of m quadrilaterals $[a_{i-1,j-1}, a_{i,j-1}, a_{i,j}, a_{i-1,j}]$ (if $i = m$, the quadrilaterals degenerate into triangles) in such a way that either:

(a) all edges of the link lie above a quadrilateral,

or

(b) all edges of the link lie below a quadrilateral,

or

(c) The quadrilateral is very small and contains a single double point corresponding to a pair of edges, one of which lies below the quadrilateral and the other above it.

By the discussion in (2) and (3) above, for each such quadrilateral we may find a point b_{ij} such that

(i) $a_{i-1,j-1}b_{ij} > 0$, $b_{ij}a_{i,j-1} > 0$

(ii) $a_{i-1,j}b_{ij} > 0$, $b_{ij}a_{ij} > 0$

(iii) $[a_{i-1,j-1}, a_{i,j-1}, a_{i,j}, a_{i-1,j}, b_{ij}]V^* = \emptyset$ or $[a_{i-1,0}, a_{i,0}]$ if $j = 0$.

It then follows that each pair of (negative) edges $a_{0,j-1}c$ and $a_{0,j}c$ "shares" a sawtooth. The bases of these sawteeth are the points

$$a_{0,j-1}, a_{1,j-1}, \dots, a_{m,j-1} = c \quad \text{on } a_{0,j-1}c$$

$$a_{0,j}, a_{1,j}, \dots, a_{m,j} = c \qquad \text{on } a_{0,j}c$$

and the common vertices are the points $b_{1j}, b_{2j}, \dots, b_{mj}$. This completes the proof of Lemma 2.3.4, for the case $\mathfrak{F} = \mathfrak{S}^{b3}_{ac}$.

Next suppose $\mathfrak{F} = \mathfrak{S}^{b5}_{ac}$. As in the case $\mathfrak{F} = \mathfrak{S}^{b3}_{ac}$, we first consider the case where ac and bc are close enough so that we can erect saw-teeth on both edges which share tips $(d_1, \dots, d_m)$. Denote the subdivisions on ac by $a = a_0, a_1, \dots, a_m = c$ and those on ab by $a = b_0, b_1, \dots, b_m = b$. These sawteeth shall be such that

$$[a_i, a_{i+1}, b_i, b_{i+1}, d_i]V^* = [a_i, a_{i+1}] \, .$$

(if ac and bc are sufficiently close this is always possible). Then $\mathfrak{F}$ can be replaced by the sequence of deformations

$$\mathfrak{F} = \left(\mathfrak{S}^{d_1 \cdots d_m}_{b_0 \cdots b_m}\right)^{-1} \mathfrak{F}_m \cdots \mathfrak{F}_2 \mathfrak{F}_1 \mathfrak{S}^{d_1 \cdots d_m}_{a_0 \cdots a_m}$$

where

$$\mathfrak{F}_i = \left(\mathfrak{R}^{a_i}_{b_i d_{i+1}}\right)^{-1} \left(\mathfrak{R}^{b_i}_{a_i d_{i+1}}\right) \quad \text{if } b_i a_i < 0 \text{ and } i \neq m$$

$$= \left(\mathfrak{R}^{a_i}_{d_i b_i}\right)^{-1} \left(\mathfrak{R}^{b_i}_{a_i d_{i+1}}\right) \quad \text{if } b_i a_i < 0 \text{ and } i \neq m$$

$$\mathfrak{F}_m = \mathfrak{R}^{b_m}_{d_m a_m} = \mathfrak{R}^{b}_{d_m c} \, .$$

The generalization to the case where ac and bc are not close proceeds by the method used in the case $\mathfrak{F} = \mathfrak{S}^{b3}_{ac}$.

The cases $(\mathfrak{S}^3)^{-1}$ and $(\mathfrak{S}^5)^{-1}$ may be handled by using the inverses of the replacement sequences for $\mathfrak{S}^3$ and $\mathfrak{S}^5$. Thus the proof of Lemma 2.3.4 is complete. ||

Proof of Lemma 2.3.5. Write V^{**} in the form $V^{**} = \mathcal{F}_2\mathcal{F}_1V^*$. We must prove the lemma for all pairs $\mathcal{F}_2\mathcal{F}_1$, where $\mathcal{F}_1$ ranges over $\{\mathcal{E}^2, \mathcal{E}^4, \mathcal{E}^6, \mathcal{E}^7, \mathcal{S}^{-1}\}$ and $\mathcal{F}_2$ ranges over the inverses of this set (cf. Table 1).[6] First observe that, from the symmetry of the hypotheses in Lemma 2.3.5, if $\mathcal{F}_2\mathcal{F}_1$ is admissible, then $\mathcal{F}_1^{-1}\mathcal{F}_2^{-1}$ is also admissible. Moreover, if there is a deformation chain satisfying the conclusions of Lemma 2.3.5 and replacing $\mathcal{F}_2\mathcal{F}_1$, then its inverse will be a replacement chain for $\mathcal{F}_1^{-1}\mathcal{F}_2^{-1}$. Therefore the cases to be considered separately are $\mathcal{F}_2\mathcal{F}_1 = \mathcal{S}\mathcal{S}^{-1}$, $\mathcal{S}\mathcal{E}$, and $\mathcal{E}^{-1}\mathcal{E}$ (since a solution for $\mathcal{S}\mathcal{E}$ implies a solution for $\mathcal{E}^{-1}\mathcal{S}^{-1}$, etc.).

Case 1: $\mathcal{F} = \mathcal{S}\mathcal{S}^{-1}$. Suppose $\mathcal{S} = \mathcal{S}^{d_1\cdots d_m}_{c_0\cdots c_m}$ and $\mathcal{S}^{-1} = \left(\mathcal{S}^{b_1\cdots b_n}_{a_0\cdots a_n}\right)^{-1}$.

If $[a_0, a_n] = [c_0, c_m]$ then by Lemma 2.3.3 the chain may be replaced by an equivalent chain of type $\mathcal{R}$ and type $\mathcal{W}$ operations, both of which leave height unaltered and so satisfy the conditions of Lemma 2.3.5.

Next consider the case $[a_0, a_n] \neq [c_0, c_m]$. Note that if $h(V^*) = h$, then $h(\hat{V}) = h+1$ and $h(V^{**}) = h$. To begin the replacement sequence, construct a sawtooth on c_0c_m which avoids $[a_0, b_1, a_1]$, thereby reducing height to $h-1$. We then apply $(\mathcal{S}^{b_1}_{a_0a_1})^{-1}$, increasing height to h. Next replace the sawtooth on c_0c_m by a second sawtooth which avoids $[a_1, b_2, a_2]$. By Lemma 2.3.3, this may be accomplished by a sequence of operations of type $\mathcal{R}$ and $\mathcal{W}$, both of which leave height fixed at h. Now construct a sawtooth on a_0a_1, say $\mathcal{S}^{\beta_1\cdots\beta_k}_{\alpha_1\cdots\alpha_k}$, to reduce the height to $h-1$, and following this, apply $(\mathcal{S}^{\beta_1\cdots\beta_k b_2}_{\alpha_1\cdots\alpha_k a_2})^{-1}$, returning the height to h. In this fashion we may remove the teeth from the sawtooth at a_0a_n, one at a time, by a sequence of operations which never increase the height to more than h. At the last step, the sawtooth on c_0c_m may be moved to its final position.

---

6 The operation $(\mathcal{E}^1)^{-1}$ is also admissible, however it may be regarded as a special case of type $\mathcal{S}^{-1}$.

Case 2: $\mathcal{F} = \mathcal{S}\mathcal{E}$. Let $\mathcal{S} = \mathcal{S}^{q_1 \cdots q_m}_{p_0 \cdots p_m}$ and $\mathcal{E} = \mathcal{E}^b_{ac}$.

If $[p_0, p_m] \neq [b, c]$ or $[a, b]$, then we may by Lemma 2.1.2 erect a sawtooth on $[p_0, p_m]$ that avoids $[a, b, c]$, say $\mathcal{S}^{b_1 \cdots d_n}_{c_0 \cdots c_n}$. We tentatively replace $\mathcal{F}$ by:

$$\mathcal{S}^{q_1 \cdots q_m}_{p_0 \cdots p_m} \left(\mathcal{S}^{d_1 \cdots d_n}_{c_0 \cdots c_n}\right)^{-1} \mathcal{E}^b_{ac}\, \mathcal{S}^{d_1 \cdots d_n}_{c_0 \cdots c_n}$$

and then use Lemma 2.3.3 to replace $\mathcal{S}^{q_1 \cdots q_m}_{p_0 \cdots p_m} (\mathcal{S}^{d_1 \cdots d_n}_{c_0 \cdots c_n})^{-1}$ by a sequence of operations of type $\mathcal{R}$ or $\mathcal{W}$. Since $h(\mathcal{S}^{d_1 \cdots d_n}_{c_0 \cdots c_n} V^*) < h(V^*)$, therefore $h(\mathcal{E}^b_{ac}\, \mathcal{S}^{d_1 \cdots d_n}_{c_0 \cdots c_n} V^*) < h(\mathcal{E}^b_{ac} V^*) = h(\hat{V})$, and since type $\mathcal{R}$ and type $\mathcal{W}$ operations leave height fixed, the conditions of Lemma 2.3.5 are satisfied. The cases $[p_0, p_m] = [b, c]$ and $[p_0, p_m] = [a, b]$ will be treated next. Each of these will involve several subcases, depending on the choice of the admissible type $\mathcal{E}$ operation. Recall that $p_0 p_m < 0$.

If $[p_0, p_m] = [b, c]$, and $k = 2$ or $6(ac > 0)$, pick a point d on ac such that $dp_i > 0$ for each $i = 0, \cdots, m-1$. Such a point exists because $cp_i > 0$ for each $i = 0, \cdots, m-1$, hence we can always find d close to c. Use the replacement sequence:

$$\mathcal{S}^{q_1 \cdots q_m}_{p_0 \cdots p_m} \mathcal{E}^b_{ac} = (\mathcal{R}^d_{ab})^{-1} \mathcal{W}^{p_0 q_1}_{d\ p_1} \cdots \mathcal{W}^{p_{m-1} q_m}_{d\ \ p_m} \mathcal{R}^d_{ac}\,.$$

(Each type-$\mathcal{W}$ operation is applicable because $\mathcal{S}^{q_1 \cdots q_m}_{p_0 \cdots p_m}$ is applicable.)

If $[p_0, p_m] = [a, b]$ and $k = 4$ or $6\,(ac > 0)$ a similar replacement sequence may be found, as follows: pick a point d on ac such that $p_i d > 0$ for each $i = 1, \cdots, m-1$. (Choose d close to a.) Then:

$$\mathcal{S}^{q_1 \cdots q_m}_{p_0 \cdots p_m} \mathcal{E}^b_{ac} = (\mathcal{E}^d_{bc})^{-1} \mathcal{W}^{q_m\ \ p_m}_{p_{m-1} d} \cdots \mathcal{W}^{q_1 p_1}_{p_0 d} \mathcal{R}^d_{ac}\,.$$

Next consider the case $[p_0, p_m] = [b, c]$ and $k = 7\,(ac < 0)$. Let $S = [p_0, q_1, p_1, \cdots, q_m, p_m]$. Note that S includes $[p_0, q_1, p_1] \cup [p_1, q_2, p_2] \cup \cdots \cup [p_{m-1}, q_m, p_m]$, and also that S meets the edge ac in precisely the single point $c = p_m$. By trivial modifications of the proof of Lemma 2.1.2, we may establish that it is possible to erect a sawtooth on the edge ac in such a way that it avoids the convex set S. Suppose that this sawtooth results from applying $\mathcal{S}^{f_1 \cdots f_n}_{e_0 \cdots e_n}$ to the edge $ac = e_0 e_n$ of V^*. Now, choose a point d on $e_n f_n$ in such a way that $dp_i > 0$ for each $i = 1, \cdots, m-1$. Such a point exists because by hypothesis $bc < 0$, hence $cb > 0$, hence $cp_i > 0$ for each $i = 1, \cdots, m-1$, hence we may find d close to c. Introduce the replacement sequence

$$\mathcal{S}^{q_1 \cdots q_m}_{p_0 \cdots p_m} \mathcal{E}^{b}_{ac} = (\mathcal{E}^{c}_{ab})^{-1} \left(\mathcal{S}^{f_1 \cdots f_n}_{e_0 \cdots e_n}\right)^{-1} \mathcal{T}^{d\ c}_{f_n b} \mathcal{W}^{p_0 q_1}_{d\ p_1} \cdots \mathcal{W}^{p_{m-1} q_m}_{d\quad p_m} \mathcal{R}^{d}_{f_n c} \mathcal{S}^{f_1 \cdots f_n}_{e_0 \cdots e_n}.$$

(In this replacement sequence, each type-$\mathcal{W}$ operation is applicable because the sawtooth on ac was constructed to avoid S.) Then replace the type-$\mathcal{T}$ operation by an appropriate product of operations of type $\mathcal{R}$ and type $\mathcal{W}$, by the method of Lemma 2.3.2.

Finally, suppose $[p_0, p_m] = [a, b]$ and $k = 7\ (ac < 0)$. As above, we may erect a sawtooth on ac by some operation of type $\mathcal{S}$, say $\mathcal{S}^{f_1 \cdots f_n}_{e_0 \cdots e_n}$, in such a way that this sawtooth avoids the set S which was defined above. Choose a point d on $e_0 f_1$ in such a way that $p_i d > 0$ for each $i = 1, \cdots, m-1$ (we may find d close to a). Then construct the replacement sequence:

$$\mathcal{S}^{q_1 \cdots q_m}_{p_0 \cdots p_m} \mathcal{E}^{b}_{ac} = (\mathcal{E}^{a}_{bc})^{-1} \left(\mathcal{S}^{f_1 \cdots f_n}_{e_0 \cdots e_n}\right)^{-1} \mathcal{T}^{da}_{cf_1} \mathcal{W}^{q_m\quad p_m}_{p_{m-1} d} \cdots \mathcal{W}^{q_1 p_1}_{p_0 d} \mathcal{R}^{d}_{af_1} \mathcal{S}^{f_1 \cdots f_n}_{e_0 \cdots e_n}.$$

Case 3: $\mathcal{F} = \mathcal{E}^{-1}\mathcal{E}$. Suppose $\mathcal{E}^{-1} = (\mathcal{E}^{e}_{df})^{-1}$ and $\mathcal{E} = (\mathcal{E}^{b}_{ac})$. Observe that $\mathcal{E}^{b}_{ac}$ produces at least one negative edge. Construct a sawtooth on one of the negative edges produced by $\mathcal{E}^{b}_{ac}$, say $\mathcal{S}^{b_1 \cdots b_m}_{a_0 \cdots a_m}$. Using it, construct the temporary replacement sequence:

$$(\mathfrak{E}^{e}_{df})^{-1}(\mathfrak{E}^{b}_{ac}) = (\mathfrak{E}^{e}_{df})^{-1}\left(\mathfrak{S}^{b_1\cdots b_m}_{a_0\cdots a_m}\right)^{-1}\left(\mathfrak{S}^{b_1\cdots b_m}_{a_0\cdots a_m}\right)(\mathfrak{E}^{b}_{ac}) .$$

This replacement sequence is of the form $(\mathfrak{E}^{-1}\mathfrak{S}^{-1})(\mathfrak{S}\mathfrak{E})$. Let $\hat{V} = \mathfrak{E}V^*$, $V_1 = \mathfrak{S}\hat{V}$, $V_2 = \mathfrak{S}^{-1}V_1$, $V^{**} = \mathfrak{E}^{-1}V_2$. Observe that $h(V^*) < h(\hat{V})$ by hypothesis and $h(V_1) < h(\hat{V})$ because an $\mathfrak{S}$ operation reduces height. Then the subsequence $V^* \to \hat{V} \to V_1$ satisfies the condition of Lemma 2.3.5, case 2, because it is type $\mathfrak{S}\mathfrak{E}$, hence we may apply the previously determined replacement sequence to this subsequence. Similarly, in the subsequence $V_1 \to V_2 \to V^{**}$ we have $h(V_1) < h(V_2) = h(\hat{V})$ and $h(V^{**}) < h(V_2)$, hence we may apply the results of case 2 a second time to the subsequence $V_1 \to V_2 \to V^{**}$. The product of the two replacement sequence so obtained then yields a replacement sequence for case 3.

This completes the proof of Lemma 2.3.5, and hence of Theorem 2.3. ||

The algebraic analogue of Theorem 2.3 now follows almost immediately. As before, we will represent a braid β by a word on the generators $\sigma_1, \cdots, \sigma_{n-1}$ of B_n. If we wish to stress the fact that a link is a closed braid, we will use the symbol $\hat{\beta}$ instead of V. If $\hat{\beta}$ is a link which is represented by the braid $\beta \in B_n$, and if we wish to stress the fact that β is a braid on n strings, we will use the symbol (β, n) instead of β.

COROLLARY 2.3.1. *Let $\hat{\beta}$ and $\hat{\beta}'$ be two closed braids, with braid representatives $(\beta, n) = (\sigma_{\mu_1}^{\epsilon_1} \cdots \sigma_{\mu_r}^{\epsilon_r}, n)$, $(\beta', n') = (\sigma_{\tau_1}^{\delta_1} \cdots \sigma_{\tau_k}^{\delta_k}, n')$. Then $\hat{\beta}$ is combinatorially equivalent to $\hat{\beta}'$*[7] *if and only if there is a deformation chain $(\beta,n) = (\beta_1,n_1) \to (\beta_2,n_2) \to \cdots \to (\beta_s,n_s) = (\beta',n')$ joining (β,n) to (β',n'), such that each closed braid (β_{i+1},n_{i+1}) in the chain can be obtained from its predecessor (β_i,n_i) by applying one of the following moves:*

---

7 The reader is reminded that closed braids which are combinatorially equivalent define the same oriented link isotopy type, in the notation of Crowell and Fox, 1963, p. 8.

$\mathfrak{M}_1$: *Replace* β_i *by any other word in* B_{n_i} *which is conjugate to* β_i. *Set* $n_{i+1} = n_i$.

$\mathfrak{M}_2$: *Replace* (β_i, n_i) *by* $(\beta_i \sigma_{n_i}^{\pm 1}, n_i + 1)$; *or, if* $\beta_i = \gamma \sigma_{n_i - 1}^{\pm 1}$, *where the braid word* γ *only involves the generators* $\sigma_1, \cdots, \sigma_{n_i - 2}$, *replace* (β_i, n_i) *by* $(\gamma, n_i - 1)$.

Proof. Let $\hat{\beta}, \hat{\beta}'$ be closed braids, which are defined by means of particular projections onto a plane E^2 which is perpendicular to the braid axis ℓ. Observe that open braids (β, n), (β', n') can be recovered from the closed braids $\hat{\beta}, \hat{\beta}'$ by cutting open along a half-plane P through the axis and reading off the braid words $\sigma_{\mu_1}^{\epsilon_1} \cdots \sigma_{\mu_r}^{\epsilon_r}$, $\sigma_{\delta_1}^{\tau_1} \cdots \sigma_{\delta_k}^{\tau_k}$ from the projections of $\hat{\beta}, \hat{\beta}'$. The integers n, n' will be the cardinality of the point sets $\beta \cap P$, $\beta' \cap P$. (Of course, the braid words so-obtained are only unique up to cyclic permutation, but n and n' are uniquely determined by the link diagram for $\hat{\beta}$ and $\hat{\beta}'$.)

If $\hat{\beta}$ and $\hat{\beta}'$ are combinatorially equivalent, there is (by Theorem 2.3) a deformation chain $\hat{\beta} = \hat{\beta}_0 \to \hat{\beta}_1 \to \hat{\beta}_2 \to \cdots \to \hat{\beta}_s = \hat{\beta}'$ joining β to $\hat{\beta}'$, with the property that each $\hat{\beta}_i$ is a closed braid, and also $\hat{\beta}_i = \mathfrak{F}_i \hat{\beta}_{i-1}$, $1 \le i \le s$, where $\mathfrak{F}_i$ is a single operation of type $\mathfrak{R}$ of type $\mathfrak{W}$.

If $\hat{\beta}_i = \mathfrak{W}_{cd}^{ab} \hat{\beta}_{i-1}$, then the projection of the tetrahedron [a, b, c, d] will in general contain double points. However, we may always alter the sequence by moving the rest of the braid away from [a, b, c, d] by operations of type $\mathfrak{R}$, then perform $\mathfrak{W}$, and afterwards restore the braid to its original position by further operations of type $\mathfrak{R}$. Hence we may assume without loss of generality that whenever $\hat{\beta}_i = \mathfrak{W}_{cd}^{ab} \hat{\beta}_{i-1}$ in the deformation chain joining $\hat{\beta}$ to $\hat{\beta}'$, the projection of the tetrahedron [a, b, c, d] is free of double points.

We now associate with each closed braid in the deformation chain a pair (β_i, n_i), where $\beta_i \in B_{n_i}$. Observe that if $\beta_i = \mathfrak{F}_i \beta_{i-1}$, and if $\mathfrak{F}_i$ is type $\mathfrak{R}$, then the string index n_i will be the same as n_{i-1}, however if $\mathfrak{F}_i$ is type $\mathfrak{W}$, then $n_i = n_{i-1} \pm 1$.

The group B_{n_i} is the fundamental group of the space $B_{0,n_i}E^2$ of Chapter 1. Hence we may associate with each braid word in B_{n_i} a homotopy equivalence class of $\tilde{z}^0$-based loops, where two loops are homotopy-equivalent if there is a homotopy (rel $\tilde{z}^0$) between them. On the other hand, we may associate with each *closed* braid an equivalence class of *freely* homotopic loops in the space $B_{0,n_i}E^2$, where two loops are (freely) homotopic if there is a homotopy between them which is *not* required to keep $\tilde{z}^0$ fixed. It then follows that we may associate with each closed braid a *conjugacy class* of elements in the group B_{n_i} [see, for example, Hu's "Homotopy Theory," p. 125, Section 14].

Now suppose that $\hat{\beta}_i = (\beta_i, n_i)$ and that $\hat{\beta}_i = \mathfrak{R}^b_{ac}\hat{\beta}_{i-1}$. Since type $\mathfrak{R}$ operations preserve string index, it follows that $n_i = n_{i-1}$ and that β_i and β_{i-1} represent conjugate elements of B_{n_i}. Hence β_i may be obtained from β_{i-1} a move of type $\mathfrak{M}_1$. If $\hat{\beta}_i = (\mathfrak{W}^{ab}_{cd})^{\pm 1}\hat{\beta}_{i-1}$, then $n_i = n_{i-1} \pm 1$. Since the tetrahedron $[a, b, c, d]$ contains no double points, the operation $(\mathfrak{W}^{ab}_{cd})^{\pm 1}$ may be realized by a single operation of type $\mathfrak{M}_2$. This completes the proof of the first part of Corollary 2.3.1. The second part of the proof is trivial. ||

Corollary 2.3.1 contains a complete translation of the question of when two links are combinatorially equivalent (as defined in Section 2.1) into an algebraic problem about the sequence of braid groups $\{B_1, B_2, B_3, \cdots\}$. This problem naturally divides into two sub-problems, the first of which (corresponding to the moves M_1) is the conjugacy problem in the braid group. Accordingly, our next step will be to discuss the solution to the conjugacy problem which was given by F. Garside in 1969. Following that, we will return (in Section 2.4) to the larger question of combinatorial equivalence of arbitrary closed braids.

2.3. *The conjugacy problems in Artin's braid group*

Our goal in this section will be to present and to extend results due to F. Garside, 1969, on the solution to the conjugacy problem in the group

B_n, i.e., to give an algorithm which enables one to decide, for an arbitrary pair of elements $\beta, \gamma \in B_n$, whether there exists an element $\delta \in B_n$ such that $\beta = \delta\gamma\delta^{-1}$.

The first part of the solution is the development, in Theorem 2.5 of a new normal form for elements in B_n. This normal form is very different from the normal form developed by Artin and presented earlier in Corollary 1.8.2, Chapter 1. To describe it, we require a definition. The group B_n will be regarded as the abstract group on the generators $\sigma_1, \cdots, \sigma_{n-1}$ with defining relations (1-1) and (1-2):

(1-1) $$\sigma_i\sigma_j = \sigma_j\sigma_i \quad \text{if} \quad |i-j| \geq 2, 1 \leq i,j \leq n-1 .$$

(1-2) $$\sigma_i\sigma_{i+1}\sigma_i = \sigma_{i+1}\sigma_i\sigma_{i+1}, \qquad 1 \leq i \leq n-2 .$$

A word in these generators is said to be a *positive* word[8] if it involves the letters σ_i $(1 \leq i \leq n-1)$, but does not involve any letter σ_i^{-1}. Of particular interest will be the positive braid word

(2-12) $$\Delta = (\sigma_1\sigma_2 \cdots \sigma_{n-1})(\sigma_1\sigma_2 \cdots \sigma_{n-2}) \cdots (\sigma_1\sigma_2)(\sigma_1) .$$

The reader is referred to Figure 11 for a picture of the braid Δ. It may be thought of as a "half-twist," accomplished by holding the top of the braid fixed, and attaching the string bottoms to a rod which is then turned over once. In Garside's treatment, each element $\beta \in B_n$ is shown to admit a unique normal form:

(2-13) $$\beta = \Delta^m P, \quad m \text{ an integer, } P \text{ a positive word.}$$

The integer m is called the *power* of β, and the word P is called the *tail* of β. Roughly speaking, m and P have the following meaning: Starting with an arbitrary word W which represents the element β, one considers all possible braid words of the form $\Delta^{-k}W$, as k ranges over

---

8 Positive braid words were also considered by [Burau, 1936] who gave them the descriptive name "gleichsinnig verdrillte Zopf."

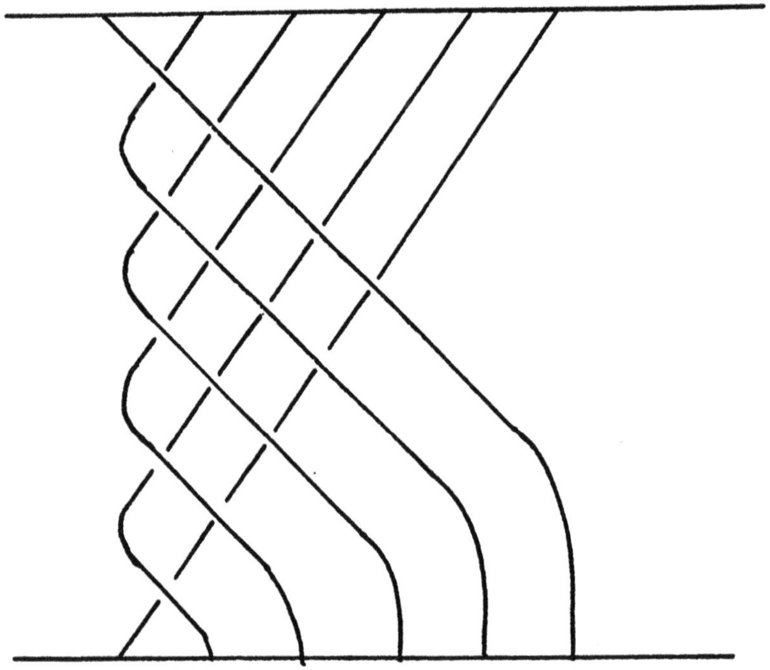

Fig. 11. The braid Δ in B_6.

the integers. Then m is the largest possible integer with the property that $\Delta^{-m}W$ is equivalent (in B_n) to a positive word, P. The particular choice of a *unique* positive word P will become clear later.

The solution to the conjugacy problem in B_n, as given in Theorem 2.6, is based on similar ideas. It is shown that one may choose a unique representative $\Delta^t T$ for the conjugacy class of an element $\beta \in B_n$. The integer t is called the *summit power* of β, and the positive word T is called the *summit tail* of β. Roughly speaking, one finds the summit power t of β by considering all braids of the form $\Delta^{-k}W$ where k ranges over the integers and W is any braid word which represents any conjugate of β in B_n; then t is the largest possible integer with the property that $\Delta^{-t}W$ is equivalent (in B_n) to a positive word, T. The particular choice of a *unique* positive word T, will become clear later.

To the reader who is familiar with R. H. Fox's concept of "congruence" of knots [Fox, 1958], we may also describe t in another way. Suppose $\beta \in B_n$ defines a *knot* $\hat{\beta}$. Then t is the largest possible integer such that $\hat{\beta}$ is congruent mod $([\frac{t}{2}], n)$[9] to a closed braid $\hat{T}$ which is represented by a positive braid word $T \in B_n$.

An algorithm for computing the power and tail of an element $\beta \in B_n$ is given in Theorem 2.5. An algorithm for computing the summit power and summit tail of an element $\beta \in B_n$ is given just before the statement of Theorem 2.6. The assertion that two elements of B_n are conjugate iff they have the same summit form as in Theorem 2.6. Following the proof of Theorem 2.6, we study Garside's results with the object of obtaining a deeper understanding of his solution. Theorem 2.7 is a new result, which may be considered as a first step in the utilization of Theorem 2.6 to solve the algebraic link problem.

Since the proof of Garside's result is quite long and complicated, and is presented in full detail in the original published version, we do not wish to simply repeat his proof. Instead, we attempt to prove the main points, omitting details of a computational nature. The interested reader should be able to work out those details for himself from the hints in the text; if not, he may refer to the original paper. Theorem 2.7, which is a new result, will be proved in detail.

For the remainder of this section, our point of view will be mainly algebraic. We remark that the geometric implications of Theorems 2.6-2.7 are not well understood, and remain a topic for future research.

We begin by noting that the defining relations (1-1) and (1-2) in the braid group B_n have the special property that no inverse of a generator appears in either relation. Hence there is an abstract semigroup S_n which admits a presentation with generators $s_1, \cdots, s_{n-1}$ and defining

---

9 The symbol $[\frac{t}{2}]$ denotes $\frac{t}{2}$ if t is even, or $\frac{t-1}{2}$ if t is odd.

relations $(1\text{-}1)_s$ and $(1\text{-}2)_s$, where $(1\text{-}1)_s$ and $(1\text{-}2)_s$ mean relations (1-1) and (1-2) with the generators $s_1, \cdots, s_{n-1}$ substituted for $\sigma_1, \cdots, \sigma_{n-1}$. Garside's basic idea is to transfer from S_n to B_n information easily obtained in S_n. The transfer is accomplished via Theorem 2.4, which asserts that the semigroup S_n embeds naturally in the group B_n. The first observation, however, is that the word problem is very easily solved in S_n:

NOTATION. Symbols V, W, X denote words in the symbols $s_1, \cdots, s_{n-1}$. *Positive equality* $V \doteq W$ means that V and W define the same element of S_n. *Identity* $V \equiv W$ means that V and W are identical words. If $W = s_{i_1} \cdots s_{i_k}$ then the symbol $\hat{W}$ will be used to denote the word $\hat{W} = s_{n-i_1} \cdots s_{n-i_k}$. The symbol $L(V)$ denotes the letter length of V. Note that if $V \doteq W$, then $L(V) = L(W)$. If $V \equiv V_0 \doteq V_1 \doteq \cdots \doteq V_r \equiv W$, and if each V_i can be obtained from V_{i-1} by a single application of one of the relations $(1\text{-}1)_s$ or $(1\text{-}2)_s$ then one says that W can be obtained from V by a *transformation of chain-length* r.

The diagram of a word solves the word problem in S_n: Let W be a word in S_n. Let $D_1(W)$ be the set of words obtainable from W by transformations of chain-length 0 or 1. Iteratively, let $D_i(W)$ be the set of words obtainable from $D_{i-1}(W)$ by transformations of chain-length 0 or 1. Since all words of $D_i(W)$ have letter length $L(W)$ and only finitely many words of S_n have letter length $L(W)$, it follows that for some integer i,

$$W \in D_1(W) \subset D_2(W) \subset \cdots \subset D_i(W) = D_{i+1}(W) .$$

Then $D_i(W) = D_j(W)$ for all $j > i$ and $D_i(W)$ is called the *diagram* $D(W)$ of the word W. Clearly $V \doteq W$ if and only if $D(V) = D(W)$. This solves the word problem in S_n.

A particular word $\Delta^{\#}$ of S_n is of special importance in the transfer of information from S_n to B_n. This word will arise in the proof of Theorem 2.4 and in both the word and conjugacy problems for B_n. Define:

$$\Delta^{\#} = (s_1 s_2 \cdots s_{n-1})(s_1 s_2 \cdots s_{n-2}) \cdots (s_1 s_2)(s_1) .$$

LEMMA 2.4.1. *The word* $\Delta^{\#}$ *in* S_n *has the following properties:*

(i) *For each word* $V \in S_n$, $\Delta^{\#} V \doteq \hat{V}\Delta^{\#}$.

(ii) *For each* $j(1 \leq j \leq n-1)$, $D(\Delta^{\#})$ *contains a word with initial letter* s_j *and a word with final letter* s_j.

Proof. One begins by showing that as a consequence of the defining relations in S_n:

$$\Delta^{\#} s_{n-i} \doteq s_i \Delta^{\#} \qquad 1 \leq i \leq n-1 .$$

Property (i) then follows by induction on $L(V)$. Property (ii) can be established by showing that

$$\Delta^{\#} = (s_j s_{j+1} \cdots s_{n-1})(s_1 s_2 \cdots s_{n-2}) \cdots (s_1 s_2)(s_1)(s_{n-1} s_{n-2} \cdots s_{n-j+1}) \qquad 2 \leq j \leq n-1$$

as a consequence of the defining relations in S_n. ||

THEOREM 2.4. *The mapping* $e : s_i \mapsto \sigma_i$ $(i = 1, 2, \cdots, n-1)$ *induces an embedding of the semigroup* S_n *in the group* B_n.

Proof. By [Clifford and Preston, 1961, pp. 34-36], it suffices to show that:

(i) S_n is right and left cancellable (i.e., in S_n, if $AX \doteq AY$, then $X \doteq Y$; and if $XA \doteq YA$, then $X \doteq Y$).

(ii) S_n is right reversible (i.e., if $X, Y \in S_n$, then there exist $U, V \in S_n$ such that $UX \doteq VY$).

By induction on the letter length of A, it suffices to prove that $AX \doteq AY$ implies $X \doteq Y$ in the case that $L(A) = 1$. This in turn follows by induction on both the chain length of the transformation taking AX to AY and also on $L(X) = L(Y)$. The induction requires simultaneous proof of the additional properties listed below:

(iii) Suppose that $s_i X \doteq s_k Y$. Then

$$\text{if} \begin{cases} i = k , & \text{it follows that } X \doteq Y \\ |i-k| \geq 2, & \text{it follows that for some } Z,\ X \doteq s_k Z,\ Y \doteq s_i Z \\ |i-k| = 1, & \text{it follows that for some } Z,\ X \doteq s_k s_i Z,\ Y \doteq s_i s_k Z . \end{cases}$$

Similarly, the proof that $XA \doteqdot YA$ implies $X \doteqdot Y$ involves proofs of the following:

(iv) Suppose that $Xs_i \doteqdot Ys_k$. Then

$$\text{if} \begin{cases} i = k \quad , & \text{it follows that } X \doteqdot Y \\ |i-k| \geq 2, & \text{it follows that for some } Z,\ X \doteqdot Zs_k,\ Y \doteqdot Zs_i \\ |i-k| = 1, & \text{it follows that for some } Z,\ X \doteqdot Zs_is_k,\ Y \doteqdot Zs_ks_i . \end{cases}$$

The assertions of (iii) and (iv) for transformations of chain length one are rather straightforward. For transformations of greater chain-length, one factors into transformations of smaller length, first applies the inductive hypothesis about chain length, then the inductive hypothesis about letter length, and checks the multitudinous possibilities. The reader is encouraged to write out two or three examples for himself; representative calculations appear in Garside's paper.

The proof of (ii) is easier: Fix $X, Y \in S_n$, $m = L(Y)$, and $Y \equiv s_iY_1$. By an induction starting with $m-1 = L(Y_1) = 0$, one may assume that there exists a $V_1 \in S_n$ such that $\Delta^{\#m-1}X \doteqdot V_1Y_1$. By Lemma 2.4.1, $\Delta^{\#} \doteqdot V_2s_i$ for some $V_2 \in S_n$. Then

$$\Delta^{\#m}X \equiv \Delta^{\#}\cdot\Delta^{\#m-1}X \doteqdot \Delta^{\#}V_1Y_1 \doteqdot \hat{V}_1\ \Delta^{\#}Y_1 \doteqdot \hat{V}_1V_2s_iY_1 \equiv \hat{V}_1V_2Y$$

(where the third equality also uses Lemma 2.4.1). This completes the induction step and establishes (ii). ||

CONVENTION. The semigroup S_n will from now on be identified with its image $e(S_n)$ in B_n. Words in the symbols $s_1 \equiv \sigma_1, \cdots, s_{n-1} \equiv \sigma_{n-1}$ which involve no inverses and thus define elements of S_n will be called *positive words*. The word $\Delta^{\#}$ in S_n will be replaced by the positive word Δ in B_n (equation 2-12). Two words W and V will be called *positively equal* (written $W \doteqdot V$) only if W and V are both positive and define the same element of S_n (equivalently, of B_n). Note that the assertion $W \doteqdot V$ implies that W and V are positive. Equality $W = V$ means that W and V define the same element of B_n; identity $W \equiv V$

means that W and V are identical words in the symbols $\sigma_1,\cdots,\sigma_{n-1}$ and their inverses. If W is the positive word $W = \sigma_{\mu_1}\sigma_{\mu_2}\cdots\sigma_{\mu_r}$, then we define $\hat{W}$ to be the positive word $\hat{W} = \sigma_{n-\mu_1}\sigma_{n-\mu_2}\cdots\sigma_{n-\mu_r}$. The following version of Lemma 2.4.1 is valid in B_n.

LEMMA 2.5.1. *For each* j $(1 \le j \le n-1)$, *there is a positive word* X_j *such that* $\sigma_j^{-1} = \Delta^{-1}X_j$. *Also,* $\Delta\sigma_j^{-1} = \sigma_{n-j}^{-1}\Delta$ *(which, together with Lemma 2.4.1, implies that* $\Delta^{-1}V = \hat{V}\Delta^{-1}$ *for every braid word* V*).*

Proof. The first statement is an immediate consequence of Lemma 2.4.1 (ii). The second is proved after the fashion of 2.4.1 (i). ‖

DEFINITION. A positive word P is said to be *prime to* Δ if no word in $D(P)$ is of the form ΔQ. Otherwise, P is said to *contain* Δ. If P_1 and P_2 are positive words of the same letter length k, with $P_1 \equiv \sigma_{\mu_1}\cdots\sigma_{\mu_k}$ and $P_2 \equiv \sigma_{\lambda_1}\cdots\sigma_{\lambda_k}$, then P_1 is defined to be *smaller* than P_2 if the number $\mu_1\cdots\mu_k$ (considered as a numerically expanded number in the number base n) is smaller than the number $\lambda_1\cdots\lambda_k$ (base n). There is clearly a unique smallest word in any diagram $D(P)$ and this word is called the *base* of $D(P)$. If P is prime to Δ, then one writes $\overline{P}$ for the base of $D(P)$. The symbol $\overline{P}$ will only be used if P is a positive word that is prime to Δ and $\overline{P}$ is the base of $D(P)$.

THEOREM 2.5 (Garside's solution to the word problem in B_n). *If* $\beta \in B_n$, *then* β *is represented by a unique word of the form* $\Delta^m\overline{P}$, *where the integers* m *and the positive word* $\overline{P}$ *are computed from any representative* $\sigma_{\mu_1}^{\epsilon_1}\cdots\sigma_{\mu_r}^{\epsilon_r}$ *of the word* β *in the following manner:*

(i) *List the positive words* $X_1,\cdots,X_{n-1}$ *whose existence is established by Lemma 2.5.1.*

(ii) *Replace every letter* $\sigma_{\mu_i}^{-1}$ *which occurs in the braid word* $\sigma_{\mu_1}^{\epsilon_1}\cdots\sigma_{\mu_r}^{\epsilon_r}$ *by* $\Delta^{-1}X_{\mu_i}$.

(iii) *Using the property* $\Delta^{-1}V = \hat{V}\Delta^{-1}$ *(Lemma 2.5.1) collect all* Δ^{-1}*'s introduced in* (ii) *at the left, so that* β *is represented by a word of the form* $\Delta^k P_0$, *where* P_0 *is positive. Note that* $k \le 0$.

(iv) *Construct* $D(P_0)$.

(v) *In* $D(P_0)$, *choose a word* $\Delta^h P$ *such that* h *is maximal. Let* $m = h+k$. *(Note that* $h \ge 0$).

(vi) *Construct* $D(P)$. *Let* $\overline{P}$ *be the base of* $D(P)$.

Proof. To prove that the representation is unique, suppose that $\beta = \Delta^m\overline{P} = \Delta^{m'}\overline{P}'$, where, say $m' \ge m$. Then $\Delta^{m-m'}\overline{P} = \overline{P}'$ is an equality of positive words and, by Theorem 2.4 $\Delta^{m-m'}\overline{P} \doteq \overline{P}'$. Since $\overline{P}'$ is prime to Δ, $m = m'$. Thus $\overline{P} \doteq \overline{P}'$ and by uniqueness of the base of a diagram of a positive word, $\overline{P} \equiv \overline{P}'$. This completes the proof of Theorem 2.5. ||

Examples:

1. Let $n = 3$, let $\beta = \sigma_1\sigma_2^{-1}\sigma_1\sigma_2^{-1}$ and note that $\sigma_2^{-1} = \Delta^{-1}\sigma_2\sigma_1$. Then

$$\beta = \sigma_1(\Delta^{-1}\sigma_2\sigma_1)\sigma_1(\Delta^{-1}\sigma_2\sigma_1) = \Delta^{-2}\sigma_1\sigma_1\sigma_2\sigma_2\sigma_2\sigma_1$$

$$D(\sigma_1^2\sigma_2^3\sigma_1) = \{\sigma_1^2\sigma_2^3\sigma_1\},$$

and $\Delta^{-2}\sigma_1^2\sigma_2^3\sigma_1$ is the representative for β.

2. Let $n = 4$, $\beta = \sigma_1^{-1}$ and note that $\sigma_1^{-1} = \Delta^{-1}\sigma_1\sigma_2\sigma_3\sigma_1\sigma_2$. Then

$$D(\sigma_1\sigma_2\sigma_3\sigma_1\sigma_2) = \{\sigma_1\sigma_2\sigma_3\sigma_1\sigma_2, \sigma_1\sigma_2\sigma_1\sigma_3\sigma_2, \sigma_2\sigma_1\sigma_2\sigma_3\sigma_2, \sigma_2\sigma_1\sigma_3\sigma_2\sigma_3, \sigma_2\sigma_3\sigma_1\sigma_2\sigma_3\},$$

and $\Delta^{-1}(\sigma_1\sigma_2\sigma_1\sigma_3\sigma_2)$ is the representative for β.

DEFINITION. The word $\Delta^m\overline{P}$ of Theorem 2.5 is called Garside's *normal form* for β. The exponent m is called the *power* of β and the word $\overline{P}$

is called the *tail* of β. If P is any word in $D(\overline{P})$, then $\Delta^m P$ is called Garside's *standard form* for β. Note that P is prime to Δ. The algebraic sum of the indices of any given word will be called its *exponent sum*. For example $(\sigma_1)^{-3}(\sigma_3)^4\sigma_2$ is of exponent sum 2. Note that the exponent sum is an invariant of conjugacy class in B_n.

The next two lemmas are easy consequences of Theorem 2.5 and motivate the solution to the conjugacy problem (Theorem 2.6).

LEMMA 2.6.1. *Let* s *and* m *be arbitrary integers. Then in* B_n *the number of words in standard form of power* $\geq$ m *and exponent sum* s *is finite.*

Proof. Suppose $\Delta^p A$ satisfies the conditions (i) $p \geq m$ and (ii) $s = pL(\Delta) + L(A)$. Then since $L(\Delta) > 0$ and $L(A) \geq 0$, (iii) $s/L(\Delta) \geq p$. There are only finitely many integers satisfying (i) and (iii). Condition (ii) shows that L(A) is determined by p, and there are only finitely many words of given length. This completes the proof. ||

LEMMA 2.6.2. *If* $\beta = \delta^{-1}\gamma\delta$ *in* B_n, *where* δ, γ *are arbitrary braid words, then there is a positive word* B *such that* $\beta = B^{-1}\gamma B$ *in* B_n.

Proof. Suppose $\delta = \Delta^p Q$, where Q is positive (see Theorem 2.5). If p is even, let $B = Q$. Then

$$B^{-1}\gamma B = Q^{-1}\Delta^{-p}\Delta^p\gamma Q = Q^{-1}\Delta^{-p}\gamma\Delta^p Q = \delta^{-1}\gamma\delta = \beta,$$

by Lemma 2.5.1. If p is odd, let $B = \Delta Q$. Then

$$B^{-1}\gamma B = Q^{-1}\Delta^{-1}(\Delta^{1-p}\Delta^{p-1})\gamma\Delta Q = Q^{-1}\Delta^{-p}\gamma\Delta^p Q = \delta^{-1}\gamma\delta = \beta$$

again by Lemma 2.5.1. ||

DEFINITION. If B is a positive word, we will call $B^{-1}\gamma B$ a *positive conjugate* of γ.

If $\Delta \doteq IF$, where $1 \le L(I) \le L(\Delta)$, then I is called an *initial route* of Δ and F is called *an associated final route*. For example, if $n = 4$, then $\Delta \doteq \sigma_1\sigma_3\sigma_2\sigma_3\sigma_1\sigma_2$, and $\sigma_1\sigma_3\sigma_2$ is an initial route with final routes $\sigma_3\sigma_1\sigma_2$ or $\sigma_1\sigma_3\sigma_2$. The initial route $\sigma_3\sigma_1\sigma_2$ has the same final routes as $\sigma_1\sigma_3\sigma_2$.

Using $D(\Delta)$, make a list of all possible initial routes. Replace each such initial route I by the base of $D(I)$. Delete repeats. The resulting set, easily obtained algorithmically, is called *the set of initial route of* Δ. For example, if $n = 3$, the set of initial route of Δ is the set $\{\sigma_1, \sigma_2, \sigma_1\sigma_2, \sigma_2\sigma_1, \sigma_1\sigma_2\sigma_1\}$. Note that the generators of B_n, i.e., $\sigma_1, \cdots, \sigma_{n-1}$, are always initial routes (Lemma 2.4.1).

We next define the *summit set* $S(\beta)$ of an element $\beta \in B_n$, by describing a procedure for computing $S(\beta)$:

(i) Construct the list of initial routes $\{I_1, \cdots, I_k\}$ of the fundamental word Δ in B_n.

(ii) Let $S_1(\beta)$ be the set of words in standard form having power at least as great as the power of β, and obtainable from β by conjugation by a single initial route I_j $(j = 1, \cdots, k)$. Iteratively, let $S_i(\beta)$ be the set of words having power at least as great as that of β and obtainable from the words of $S_{i-1}(\beta)$ by conjugation by a single initial route of Δ.

(iii) By Lemma 2.6.1, the number of words obtainable by the process is finite so that, for some j, $S_j(\beta) = S_{j+1}(\beta)$ is the subset of $S_j(\beta)$ consisting of those elements of maximum power in $S_j(\beta)$.

Finally, define the *summit power* t of $\beta \in B_n$ to be the (common) power of the elements $S(\beta)$; the *summit tail* T of β to be the tail of the unique element in $S(\beta)$ with smallest tail; and the *summit form* of β to be the word $\Delta^t T \in B_n$.

THEOREM 2.6. *Two elements* $\beta_1, \beta_2 \in B_n$ *are conjugate if and only if they have the same summit forms.*

Proof. It is necessary to establish that the integer t and the positive word T which are computed by the algorithm given above are uniquely determined by and uniquely determine the conjugacy class of the word β. The proof begins with a definition and several lemmas.

A positive word V is said to *begin with* σ_i if D(V) contains a word of the form $\sigma_i A$, and to *end in* σ_i if D(V) contains a word of the form $A\sigma_i$. For example, the word $V = \sigma_1\sigma_2\sigma_3\sigma_1$ has diagram $D(\sigma_1\sigma_2\sigma_3\sigma_1) = \{\sigma_1\sigma_2\sigma_3\sigma_1, \sigma_1\sigma_2\sigma_1\sigma_3, \sigma_2\sigma_1\sigma_2\sigma_3\}$ hence V begins with σ_1 and σ_2 and ends with σ_1 and σ_3. The lemma which follows summarizes the properties of Δ which will be used in the proof of Theorem 2.6.

LEMMA 2.6.3.

1. *If* $i \in \{1,\cdots,n-1\}$ *then* $\Delta\sigma_i \doteq \sigma_{n-i}\Delta$. *Hence, if* V *is any positive word, then* $\Delta V \doteq \hat{V}\Delta$.
2. *If* $i \in \{1,\cdots,n-1\}$, *then there are positive words* X_i, Y_i *such that* $\Delta \doteq \sigma_i X_i \doteq Y_i \sigma_i$.
3. *If* $\Delta \doteq IF$, *then either* I *ends in* σ_i *or* F *begins with* σ_i, *for each* $i \in \{1,\cdots,n\}$, *but never both.*
4. *If* $W \doteq \Delta V$, *and* $W \doteq AB$, *then either* A *ends in* σ_i *or* B *begins with* σ_i *for each* $i \in \{1,\cdots,n-1\}$.
5. *If* W *is a positive word in* B_n, *and if* W *begins with (ends with)* σ_i *for each* $i \in \{1,\cdots,n-1\}$ *then* $W \doteq \Delta V$ *for some positive word* V.

Proof of Lemma 2.6.3. Properties 1 and 2 come from Lemma 2.5.1. Property 5 is proved by a lengthy induction and involves all of the properties of Δ already developed. Property 4 is proved by induction on L(A) and uses Properties (iii) and (iv) from the proof of Theorem 2.4. Property 4 implies all but the last assertion of Property 3, which assertion follows from Properties (iii), (iv), Theorem 2.4 and Property 5. The reader is referred to Garside's paper for details. ||

NOTATION. The symbol $\tilde{Q}$ will be used to denote a positive word which is equal to either Q or $\hat{Q}$.

LEMMA 2.6.4. *Suppose that* $\beta = \Delta^m P$, *and suppose further that for some positive word* Q, *the word* $Q^{-1}\Delta^m PQ$ *has power* r *greater than* m. *Then* PQ *contains* Δ.

Proof of Lemma 2.6.4. By hypothesis, $Q^{-1}\Delta^m PQ = \Delta^r R$, with $r > m$. Using Lemma 2.5.1, we then find:

$$\Delta^m PQ = Q\Delta^r R = \Delta^r \tilde{Q} R$$

$$PQ = \Delta^{r-m}\tilde{Q}R, \quad r-m > 0 .$$

By Theorem 2.4, it then follows that $PQ \mathrel{\dot{=}} \Delta^{r-m}\tilde{Q}R$. This completes the proof. ||

LEMMA 2.6.5. *In* B_n, *suppose* (i) *that* $\gamma = \Delta^t T$ *is in the summit set* $S(\beta)$ *of* β, (ii) *that* X *is a positive word such that* $X^{-1}\gamma X = \Delta^t Q$, *and* (iii) *that* $X = IY$ *where* I *is an initial route of maximum length. Then* $I^{-1}\gamma I$, *reduced to standard form, is in* $S(\beta)$.

Proof of Lemma 2.6.5. Since $\gamma \in S(\gamma) \subset S(\beta)$, $I^{-1}\gamma I$ has power $\leq t$. It remains to be shown that $I^{-1}\gamma I$ also has power $\geq t$.

Case 1. If $X \mathrel{\dot{=}} \Delta Y$, then $\Delta^{-1}\gamma\Delta = \Delta^{-1}(\Delta^t T)\Delta = \Delta^t\hat{T}$ has power t.

Case 2. If $L(I) < L(\Delta)$, let F be a positive word such that $IF \mathrel{\dot{=}} \Delta$. There is a partition of the indices $1,\cdots,n-1$ into two sets α and δ such that, for each $j \in \alpha$, there is a positive word U_j such that $I \mathrel{\dot{=}} U_j\sigma_j$ and, for each $j \in \delta$, there is no such word. Then

$$I^{-1}\Delta^t = I^{-1}\Delta\Delta^{t-1} = F\Delta^{t-1} = \Delta^{t-1}\tilde{F}$$

and consequently

$$\Delta^t Q = X^{-1}\gamma X = Y^{-1}I^{-1}(\Delta^t T)IY = Y^{-1}[\Delta^{t-1}\tilde{F}TI]\,Y .$$

By Lemma 2.6.4, $\tilde{F}TIY$ contains Δ. By Lemma 2.6.3 (Property 4), either $\tilde{F}TI \doteq V_j\sigma_j$ for some V_j or $Y \doteq \sigma_j Y_j$ for some Y_j for each $j \in \{1, \cdots, n-1\}$. But if the latter were true for some $j \in \delta$, then $X \doteq IY \doteq I\sigma_j Y_j$ and $I\sigma_j$ would be by Lemma 2.6.3 (Property 3) positively equal to an initial route of Δ of length greater than $L(I)$, a contradiction. Thus $\tilde{F}TI \doteq V_j\sigma_j$ for each $j \in \delta$ by the contradiction just established. Also, $\tilde{F}TI \doteq \tilde{F}TU_j\sigma_j$ for each $j \in \alpha$ by definition of α. Hence $\tilde{F}TI$ contains Δ by Property 5 of Lemma 2.6.3. This shows that $I^{-1}\gamma I = \Delta^{t-1}\tilde{F}TI$ has power $\geq t$ as desired. This completes the proof. ||

We are now ready to prove that in B_n, β_1 is conjugate to β_2 if and only if $S(\beta_1) = S(\beta_2)$.

Proof. If there is an $X \in S(\beta_1) \cap S(\beta_2)$, then X is conjugate to both β_1 and β_2; hence β_1 and β_2 are conjugate.

Suppose conversely that β_1 and β_2 are conjugate, that $\Delta^p P \in S(\beta_1)$, and $\Delta^q Q \in S(\beta_2)$, say with $p \leq q$. By Lemma 2.6.2, there is a positive word X such that $X^{-1}(\Delta^p P)X = \Delta^q Q$. If $L(X) > 0$, then $X \doteq IY$ where I is an initial route of Δ of maximum possible length $L(I) > 0$ (recall that each σ_i is an initial route of Δ). Note that $p \leq q$, so that by Lemma 2.6.5, $I^{-1}(\Delta^p P)I$ when reduced to normal form is in $S(\beta_1)$. By induction on $L(X)$, $\Delta^q Q = Y^{-1}[I^{-1}(\Delta^p P)I]Y$ has normal form in $S(\beta_1)$. Thus $S(\beta_2) \subset S(\beta_1)$ and $p = q$. Since $p = q$, the argument may be repeated with the roles of β_1 and β_2 interchanged to show that $S(\beta_1) \subset S(\beta_2)$. This completes the proof of Theorem 2.6. ||

To proceed further, we would like to examine the summit set $S(\beta)$ of an element $\beta \in B_n$. We begin with certain observations about $S(\beta)$ which are implied by the procedure for computing $S(\beta)$.

In computing the summit set of an element $\beta = \Delta^m P$ by the procedure described above, we include in the set $S_i(\beta)$ $(i = 1, 2, \cdots, k)$ all words of power $m, m+1, m+2, \cdots$ until finally the complete list of words of maximum power t is obtained. All words of power $< t$ are then ignored. In the process we must at some stage reach a first word of power $m+1$, say β_1. Since β_1 is conjugate to β, it follows that the summit sets of β and β_1 coincide. But then this implies that all words of power $m+i-1$ can be ignored, once any *single* word of power $m+i$ is obtained. Moreover, the entire summit set of an element β can be computed unambiguously from any summit representative of β. Finally, given any element $\beta = \Delta^m P$, it is possible to reach at least one summit representative $\Delta^t T_j$ of β by a sequence of words $\Delta^{m+1}P_1, \Delta^{m+2}P_2, \cdots, \Delta^t T_j$ which have the property that each $\Delta^{m+i}P_i$ is conjugate to $\Delta^m P$, and has larger power than its predecessor. These remarks, which are implicit in Theorem 2.6 and in the procedure for calculating $S(\beta)$, will prove to be extremely important in the discussion which follows below.

From this point on, our results are new, and go beyond those in Garside's paper. We recall that our motivation in studying the conjugacy problem in B_n was a wish to study the "Algebraic Link Problem," as formulated in Corollary 2.3.1. With the end in mind, one is not really so much interested in being able to reach a decision about whether particular elements in B_n are conjugate, but rather one seeks insight into the properties of a braid which cause it to fall into one conjugacy class rather than into another. Thus one might ask the question: given a braid $\beta = \Delta^m P$, in normal form, how can we tell, by inspection of the word P, whether it is a summit form? With this goal in mind,[10] we will prove:

THEOREM 2.7. *Let* $\beta = \Delta^m P$ *be in standard form in* B_n. *Then* β *has summit power* $t > m$ *if and only if* $D(P)$ *contains a word of the form*

$$\text{FRI} \ (\textit{if } m \textit{ is even}) \quad \textit{or} \quad \text{FR}\hat{\text{I}} \ (\textit{if } m \textit{ is odd}) \tag{2-14}$$

[10] Our results will *also* give a simplified algorithm for deciding whether an arbitrary pair of elements of B_n are conjugate.

for some pair of positive words I, F *such that* $\Delta \doteq IF$. *Moreover, if this happens, then* β *is conjugate to* $\Delta^{m+1}R$. *Thus* β *is a summit form if and only if* $D(P)$ *contains no words of the form* (2-14).

Before proving Theorem 2.7, we note that it has as an immediate dividend a simplified algorithm for the computation of the summit set $S(\beta)$ of a element $\beta \in B_n$:

COROLLARY 2.7.1. *The following procedure may be used to compute the summit set* $S(\beta)$, $\beta \in B_n$:

(i) *Construct a master list of all possible pairs* $\{(I_1, F_1), \cdots, (I_s, F_s)\}$ *of initial and final routes, where* $\Delta \doteq I_1 F_1 \doteq \cdots I_s F_s$.

(ii) *Suppose* $\beta = \Delta^m P$. *Construct* $D(P)$. *As each word is entered in* $D(P)$, *search the list of pairs* (I_j, F_j) *to see if there is a pair* (I, F) *such that* $P \doteq IP_1\hat{F}$ *(if* m *is even) or* $P \doteq \hat{I}P_1\hat{F}$ *(if* m *is odd). If so, replace* $\Delta^m P$ *by* $\Delta^{m+1}P_1$, *and repeat step (ii).*

(iii) *Ultimately, an element* $\beta' = \Delta^t T_0$ *will be obtained which is conjugate to* β *and which has the property that* $D(T_0)$ *fails the test in (ii) above. Then* $\beta' \in S(\beta) = S(\beta')$, *and we may compute the set* $S(\beta)$ *by the procedure outlined in steps (iv) and (v) below.*

(iv) *The following words may be entered in the list* $S(\beta)$ *without further computation, once a single entry* $\Delta^t T_j$ *has been entered:*

(a) *For each word* $\Delta^t T_j$ *on the list, include all words* $\Delta^t T_k$ *where* $T_k \in D(T_j)$.

(b) *For each word* $\Delta^t T_j$, *and for every possible partition* $T_j \doteq T_{j_1} T_{j_2}$ *of* T_j *we may enter* $\Delta^t T_{j_2} T_{j_1}$ *if* t *is even;* $\Delta^t \hat{T}_{j_2} T_{j_1}$ *if* t *is odd.*

(c) *If* $S(\beta)$ *contains any word* $\Delta^t T_j$ *which has the property that* T_j *does not involve the braid generators* $\sigma_m, \cdots, \sigma_{n-1}$, *i.e.* $T_j = T_j(\sigma_1, \cdots, \sigma_{m-1})$, *then we may enter the words* $T_j(\sigma_2, \cdots, \sigma_m), \cdots, T_j(\sigma_{1+n-m}, \cdots, \sigma_{n-1})$.

(d) *For each word* $\Delta^t T_j$ *on the list, we may enter the word* $\Delta^t \hat{T}_j$.

(v) *Let* $S(\beta)$ *be the (possibly incomplete) list developed in step (iv) above. For each word* $\Delta^t T_j \in S(\beta)$ *verify whether* $I_i^{-1}\Delta^t T_j I_i$ *has power* t, *and if so, whether it is in* $S(\beta)$, *for each initial route* I_i. *If it is not already in* $S(\beta)$, *then augment* $S(\beta)$ *by this new word and repeat step (iv) above for this new word. The procedure will end when no new words are obtained by step (v).*

We conjecture that no new words will ever be obtained from step (v) above, however we were unable to prove this (see Problem 12 in the Appendix).

REMARK. To the reader who is unfamiliar with Garside's work, our "improved" algorithm may not seem to be an improvement, however we assure the reader that it reduces the computational work considerably! Nevertheless, the algorithm may prove extremely difficult to apply, mainly because of the large number of words in the set $D(\Delta)$. If $n = 3$, $D(\Delta)$ contains 2 words; if $n = 4$, $D(\Delta)$ has length 16; R. S. D. Thomas informs me that for $n = 5$, $D(\Delta)$ has length 768, and that for $n = 6$, $D(\Delta)$ has length 292, 864! (The study of patterns in $D(\Delta)$ is, incidentally, a fascinating pastime in its own right! See [R. S. D. Thomas, 1974], for such a study.)

The remainder of this section will be occupied with the proof of Theorem 2.7. The reader who is more interested in applications than in a deeper understanding of Garside's solution may wish to skip to Section 2.4.

Proof of Theorem 2.7. The sufficiency of the condition of Theorem 2.7 is immediate. If m is even, and if $\beta = \Delta^m FRI$, then $I\beta I^{-1} = I\Delta^m FR = \Delta^m IFR = \Delta^{m+1}R$. If m is odd, and if $\beta = \Delta^m FR\hat{I}$, then $\hat{I}\beta\hat{I}^{-1} = \hat{I}\Delta^m FR = \Delta^m IFR = \Delta^{m+1}R$.

To establish necessity, we begin with several easy lemmas, followed by several others of increasing difficulty.

LEMMA 2.7.1.

$$\Delta \doteq \hat{\Delta} \tag{2-15}$$

$$\textit{if}\ U \doteq V,\ \textit{then}\ \hat{U} \doteq \hat{V}. \tag{2-16}$$

Proof of Lemma 2.7.1. These are immediate consequences of Lemma 2.6.3, Property 1. ||

LEMMA 2.7.2. *If* $\Delta \doteq IF$, *then also*

$$\Delta \doteq F\hat{I} \doteq \hat{I}\hat{F} \doteq \hat{F}I\,. \tag{2-17}$$

Proof of Lemma 2.7.2. Since $\Delta \doteq IF$, therefore $I^{-1}\Delta\hat{I} = \Delta \doteq F\hat{I}$. Conjugating IF and $F\hat{I}$ by Δ gives the remaining cases. ||

LEMMA 2.7.3. *Let* $\beta = \Delta^m P$ *be in normal form. Let* I *be an initial route. Suppose that* $I^{-1}\beta I = \Delta^q Q$, *where* $q > m$. *Then* $q = m+1$.

Proof of Lemma 2.7.3. Suppose that $q = m+r$, where $r \geq 1$. Then

$$\Delta^m PI \doteq I\Delta^{m+r}Q \doteq I\Delta^{m+r-1}\hat{Q}\Delta\,. \tag{2-18}$$

Since I is an initial route, therefore $\Delta \doteq IF$, hence by Lemma 2.7.2, $\Delta \doteq \hat{F}I$, hence

$$\begin{aligned} \Delta^m P &\doteq I\Delta^{m+r-1}\hat{Q}\hat{F} \\ P &\doteq \Delta^{r-1}\tilde{I}\hat{Q}\hat{F} \end{aligned} \tag{2-19}$$

(where $\tilde{I}$ denotes either I or $\hat{I}$, depending on whether $m+r-1$ is even or odd). But $\Delta^m P$ was by hypothesis in standard form, hence P is prime to Δ, hence r can only be 1. ||

LEMMA 2.7.4. *Let* $\beta = \Delta^m P$ *be in normal form. Then there exists an initial route* I *such that* $I^{-1}\beta I = \Delta^{m+1}Q$ *if and only if* $D(P)$ *contains a word of the form* $I\hat{Q}\hat{F}$ *(if* m *is even) or* $\hat{I}\hat{Q}\hat{F}$ *(if* m *is odd), where* F *is the final route belonging to the initial route* I.

Proof of Lemma 2.7.4. We establish sufficiency first. Suppose that $P \doteq \hat{I}\hat{Q}F$ (if m is even), or $\hat{\hat{I}}\hat{Q}F$ (if m is odd). Then

$$I^{-1}\Delta^m PI = \Delta^m \hat{Q}\hat{F}I = \Delta^m \hat{Q}\Delta = \Delta^{m+1}Q\,. \tag{2-20}$$

To prove necessity, suppose that there exists an initial route I such that

$$\begin{aligned} I^{-1}\Delta^m PI &= \Delta^{m+1}Q \\ \Delta^m PI &= I\Delta^m \hat{Q}\Delta\,. \end{aligned} \tag{2-21}$$

Since $\Delta = IF = \hat{F}I$ (by Lemma 2.7.2) it follows that

$$\begin{aligned} P &\doteq \hat{I}\hat{Q}F \quad \text{if} \quad m \text{ is even} \\ &\doteq \hat{\hat{I}}\hat{Q}F \quad \text{if} \quad m \text{ is odd}\,. \end{aligned} \tag{2-22}$$

This proves Lemma 2.7.4. ||

To motivate what follows: to compute the summit set of $\beta = \Delta^m P$, it is necessary to first compute all positive conjugates of β by a single initial route. One then lists those which have power $m+1$, or, if none exist which have power $m+1$, one lists those which have power m. According to Lemma 2.7.4, we can recognize the situation where positive conjugation by a single initial route will result in a word of power $m+1$ by inspecting the list of words in D(P). We now ask a more complicated question: how can we recognize the situation where positive conjugation by a single initial route I results in a word of power $\leq m$, but positive conjugation by a suitable sequence of initial routes $I_1 I_2 \cdots I_j$ results in a word of power m if $1 \leq j \leq k-1$, and a word of power $m+1$ if $j = k$? We begin to investigate this question in the next lemma.

LEMMA 2.7.5. *Suppose that* $\beta = \Delta^m P$ *is in standard form. Let* Z *be a positive word in* B_n. *Let* $Z \doteq IY$, *where* I *is an initial route of maximal length in* Z. *Suppose that* $Z^{-1}\beta Z$ *has power* m. *Then* $I^{-1}\beta I$ *also has power* m.

Proof of Lemma 2.7.5. If $I \equiv \Delta$, the proof is trivial, since $\Delta^{-1}\Delta^m P\Delta = \Delta^m\hat{P}$ obviously has power m.

If $I \neq \Delta$, then $\Delta \doteq IF$. Suppose that I ends in the braid generators $\eta_1,\cdots,\eta_s$ (but not in any other braid generator σ_j). Let $\tau_1,\cdots,\tau_{n-1-s}$ be the remaining braid generators. Since $\Delta \doteq IF$, it follows from Lemma 2.6.3 (Property 3) that F must begin with $\tau_1,\cdots,\tau_{n-1-s}$. Hence Y cannot begin with $\tau_1,\cdots,\tau_{n-1-s}$, for if it did, then I would not be a *maximal* initial route. By hypothesis:

$$Y^{-1}I^{-1}\Delta^m PIY = \Delta^m Q, \text{ with } Q \text{ prime to } \Delta.$$
$$\Delta^m PIY = IY\Delta^m Q .$$

Since $\Delta \doteq IF$, this gives:

$$F\Delta^{m-1}PIY = Y\Delta^m Q$$
$$\hat{F}PIY = \Delta YQ \quad \text{if} \quad m \text{ is even}$$
$$FPIY = \Delta\hat{Y}Q \quad \text{if} \quad m \text{ is odd}.$$

First treat the case where m is even. Applying Lemma 2.6.3, Property 4, with $A = \hat{F}PI$, $B = Y$, and $V = YQ$, we see that for each $i \in \{1,\cdots,n-1\}$ it must be true that either $\hat{F}PI$ ends in σ_i, or Y begins with σ_i. By hypothesis, I ends with $\eta_1,\cdots,\eta_s$, hence surely $\hat{F}PI$ ends in $\eta_1,\cdots,\eta_s$. Also, as demonstrated above, Y cannot begin with any of the remaining braid generators $\tau_1,\cdots,\tau_{n-1-s}$. Hence the only possibility is that $\hat{F}PI$ also ends in $\tau_1,\cdots,\tau_{n-1-s}$. But then, $\hat{F}PI$ ends in every braid generator $\sigma_1,\cdots,\sigma_{n-1}$. Hence, by Lemma 2.6.3, Property 5, there exists a positive word $R \in B_n$ such that

$$\hat{F}PI \doteq \Delta R \quad (\text{if } m \text{ is even})$$

Similar reasoning in the case where m is odd yields:

$$FPI \doteq \Delta R \quad (\text{if } m \text{ is odd}).$$

Now use the fact that $\Delta \doteq IF = \hat{F}I \doteq F\hat{I}$ (Lemma 2.7.2). Then

$$PI \doteq IR \quad (\text{if } m \text{ is even})$$
$$PI \doteq IR \quad (\text{if } m \text{ is odd}).$$

But then

$$I^{-1}\Delta^m PI = \Delta^m I^{-1} IR = \Delta^m R \quad (\text{if } m \text{ is even})$$
$$= \Delta^m \hat{I}^{-1}\hat{I}R = \Delta^m R \quad (\text{if } m \text{ is odd}).$$

This completes the proof. ||

To continue, we need a new concept, the "cyclic diagram" $C(\Delta^m P)$ of an element $\beta = \Delta^m P$. If $P \doteq AB$, then the positive word BA will be referred to as a *cyclic permutation* of P, and the positive word $B\hat{A}$ will be referred to as a *reflected cyclic permutation* of P. The *cyclic diagram* $C(\Delta^m P)$ of $\Delta^m P$ is a list of positive words in B_n which is defined by the rule:

1. The list $C(\Delta^m P)$ includes the word P.
2. For every word $P_j \in C(\Delta^m P)$, the list also includes every word in $D(P_j)$.
3. For every word $P_j \in C(\Delta^m P)$, the list $C(\Delta^m P)$ also includes all cyclic permutations of P_j (if m is even), *or* all reflected cyclic permutations of P_j (if m is odd).

For example, let $P = \sigma_1\sigma_3\sigma_1 \in B_4$. Then

$$C(\Delta^m P) = \{\sigma_1\sigma_3\sigma_1, \sigma_3\sigma_1^2, \sigma_1^2\sigma_3\} \text{ if } m \text{ is even}$$

$$C(\Delta^m P) = \{\sigma_1\sigma_3\sigma_1, \sigma_3\sigma_1^2, \sigma_1^2\sigma_3, \sigma_3\sigma_1\sigma_3, \sigma_3^2\sigma_1, \sigma_1\sigma_3^2, \sigma_3^3, \sigma_1^3\} \text{ if } m \text{ is odd}.$$

Several properties of $C(\Delta^m P)$ follow immediately from the definition.

LEMMA 2.7.6. *The cyclic diagram* $C(\Delta^m P)$ *has the following properties:*

1. *If* $P_j \in C(\Delta^m P)$, *then* $C(\Delta^m P_j) = C(\Delta^m P)$.
2. $C(\Delta^m \hat{P})$ *is obtained from* $C(\Delta^m P)$ *by replacing each word* P_j *in* $C(\Delta^m P)$ *by* $\hat{P}_j$.
3. *If* $m' = m \pmod 2$, *then* $C(\Delta^m P) = C(\Delta^{m'} P)$.

Proof of Lemma 2.7.6. Immediate. ||

LEMMA 2.7.7. *Let* $\Delta^m P$ *be in standard form in* B_n. *Let* X *be a positive word in* B_n. *Write* X *in the form* $X \doteq IZ$, *where* I *is an initial route of maximal possible length, and assume that* Z *is not the empty word. Suppose that* $I^{-1}\Delta^m PI$ *has power* m, *but that* $X^{-1}\Delta^m PX$ *has power* $m+1$, *with*

$$I^{-1}\Delta^m PI = \Delta^m S .$$

Then the positive word S *is in the cyclic diagram* $C(\Delta^m \hat{P})$.

Proof. Suppose that I ends in the braid generators $\eta_1,\cdots,\eta_s$ (but not in any other braid generator σ_j). Let $\tau_1,\cdots,\tau_{n-1-s}$ be the remaining braid generators. Since $\Delta \doteq IF$, it follows from Lemma 2.6.3, Property 4 that F must begin with $\tau_1,\cdots,\tau_{n-1-s}$. Hence Z cannot begin with $\tau_1,\cdots,\tau_{n-1-s}$, for if it did, then I would not be a *maximal* initial route. Now, $X^{-1}\Delta^m PX$ has power $m+1$, hence there is a positive Q such that

$$Z^{-1}I^{-1}\Delta^m PIZ = \Delta^{m+1}Q ,$$

$$\Delta^m PIZ = IZ\Delta^{m+1}Q ,$$

$$PIZ = \Delta\tilde{I}\tilde{Z}Q .$$

Applying Lemma 2.6.3, Property 4 (with $V = \tilde{I}\tilde{Z}Q$, $A = PI$, $B = Z$) it follows that for each $i \in \{1,\cdots,n-1\}$ either PI ends in σ_i or Z begins with σ_i. Now, by hypothesis, I ends in $\eta_1,\cdots,\eta_s$, hence surely PI ends in $\eta_1,\cdots,\eta_s$. Also, as demonstrated above, Z cannot begin with the remaining braid generators $\tau_1,\cdots,\tau_{n-1-s}$. Therefore PI must also end in $\tau_1,\cdots,\tau_{n-1-s}$. But then PI ends in every braid generator $\sigma_1,\cdots,\sigma_{n-1}$. Hence by Lemma 2.6.3, Property 5 there is a positive word R in B_n such that

$$PI \doteq \Delta\hat{R} \doteq R\Delta .$$

But $\Delta \doteq IF$, hence, by Lemma 2.7.2, $\Delta = \hat{F}I$, hence $PI \doteq R\hat{F}I$, or

$$P \doteq R\hat{F} . \tag{2-23}$$

Now consider $I^{-1}\Delta^m PI$:

$$I^{-1}\Delta^m PI = I^{-1}IF\Delta^{m-1}R\hat{F}I = F\Delta^m\hat{R}$$

$$\text{(2-24)} \qquad I^{-1}\Delta^m PI = \Delta^m F\hat{R} \text{ (if } m \text{ is even) or } \Delta^m \hat{F}\hat{R} \text{ (if } m \text{ is odd)}.$$

Now, in equation (2-24) the positive word $F\hat{R}$ (or $\hat{F}\hat{R}$) is necessarily prime to Δ, since it was assumed in the statement of Lemma 2.7.7 that $I^{-1}\Delta^m PI$ has power m. Let $S = F\hat{R}$ (if m is even), or $\hat{F}\hat{R}$ (if m is odd). Comparing equations (2-23) and (2-24) we see that $S \in C(\Delta^m \hat{P})$, hence Lemma 2.7.7 is true. ||

LEMMA 2.7.8 [Garside, 1969]. *Let* Q *be a positive word. Suppose that* $\Delta Q \,=.\, AW$. *Then there exist initial and final routes* I, F *(with* $\Delta \doteq IF$*) and positive words* A_0, W_0, *such that* $A \,=.\, A_0 I$ *and* $W \,=.\, FW_0$.

Proof of Lemma 2.7.8. See Theorem 12, Appendix III of Garside's paper. ||

We are now ready to establish Theorem 2.7. First, we show that the condition of Theorem 2.7 is sufficient. If $P \,=.\, FRI$ and m is even, then $(\Delta^{-1}F^{-1})(\Delta^m P)(F\Delta) = \Delta^{-1}F^{-1}\Delta^m FRIF\Delta = \Delta^{m-1}RIF\Delta = \Delta^{m+1}R$. On the other hand, if $P = FR\hat{I}$ and m is odd, then $(\Delta^{-1}\hat{F}^{-1})(\Delta^m P)(\hat{F}\Delta) = \Delta^{-1}\hat{F}^{-1}\Delta^m FR\hat{I}\hat{F}\Delta = \Delta^{m-1}R\hat{I}\hat{F}\Delta = \Delta^{m+1}R$.

To prove necessity, we will establish a slightly different result: that β has summit power $> m$ only if $C(\Delta^m P)$ contains a word of the form ΔQ. First, we show that this is equivalent to the assertion of Theorem 2.7, by showing that a necessary condition for $C(\Delta^m P)$ to contain a word ΔQ is the existence of initial and final routes I, F such that $D(P)$ contains a word of the form given in equation (2-14). To establish this, observe that if $\Delta Q \in C(\Delta^m P)$, then from the definition of $C(\Delta^m P)$ it must be true that either $\Delta Q \in D(P)$ (however since $\Delta^m P$ is in standard form, so that P is prime to Δ, this is impossible), or else $\Delta Q \doteq AW$ and $P \doteq WA$ (if m is even) or $W\hat{A}$ (if m is odd) for some pair W, A such that $L(W) \geq 1$, $L(A) \geq 1$. We may then apply Lemma 2.7.8 to conclude that there exists a pair of initial and final routes I, F, and positive words A_0, W_0 such that $A \doteq A_0 I$ and $W \doteq FW_0$. But then $P \doteq WA \doteq FW_0A_0I$ (if m is even) or $P \doteq W\hat{A} \doteq FW_0\hat{A}_0\hat{I}$ (if m is odd), and this is exactly the assertion of Theorem 2.7.

It remains to prove that β has summit power $> m$ only if $C(\Delta^m P)$ contains a word which contains Δ. Assume that β has summit power $> m$. Then, by Theorem 2.6, there is a positive word X, which is a product of initial routes, such that

$$X^{-1}\Delta^m PX = \Delta^{m+1}Q.$$

If X is an initial route, say $X = I$, where $\Delta \doteq IF$, then by Lemma 2.7.4 the diagram $D(P)$ must contain a word of the form $I\hat{Q}\hat{F}$ (if m is even) or $\hat{I}\hat{Q}\hat{F}$ (if m is odd). But then $C(\Delta^m P)$ contains the word $\hat{Q}\hat{F}I = \hat{Q}\Delta = \Delta Q$, as required.

If X is *not* an initial route, then X can always be written in the form $X \doteq I_0Z_0$, where I_0 is an initial route of maximal length, and Z_0 is not the empty word. If $I_0^{-1}\Delta^m PI_0$ has power $m+1$, then we can replace X by I_0 and apply the argument given above to conclude that $C(\Delta^m P)$ contains the word ΔQ, hence we need only consider the case where $I_0^{-1}\Delta^m PI_0$ has power $\leq m$.

Since $Z_0 \neq 1$, we can write $Z_0 \doteq I_1Z_1, Z_1 \doteq I_2Z_2, \cdots$, where $I_1, I_2, \cdots$ are successively defined as initial routes of maximal length, and $Z_1, Z_2, \cdots$ are positive words of steadily decreasing lengths, so that after say, q, steps, we obtain:

$$Z_0 = I_1I_2\cdots I_q.$$

We now have the situation:

$$\beta \text{ has power } m$$

$$I_0^{-1}\beta I_0 \text{ has power } \leq m$$

$$I_q^{-1}\cdots I_2^{-1}I_1^{-1}I_0^{-1}\beta I_0I_1I_2\cdots I_q \text{ has power } m+1.$$

Since (by Lemma 2.7.3) conjugation by a single initial route can increase the power of a word by at most 1, there must be a smallest integer k, where $1 \leq k \leq q$, with the property that

$$I_{k-1}^{-1} \cdots I_0^{-1} \beta I_0 \cdots I_{k-1} \text{ has power } m$$

(2-25)

$$I_k^{-1} \cdots I_0^{-1} \beta I_0 \cdots I_k \text{ has power } m+1 .$$

Now observe that I_0 is a maximal initial route in the positive word $I_0 \cdots I_{k-1}$, and that β has power m. Hence we can apply Lemma 2.7.5, and conclude that $I_0^{-1}\beta I_0$ must have power precisely m. Again, I_1 is a maximal initial route in the positive word $I_1 \cdots I_{k-1}$, so that Lemma 2.7.5 can be applied a second time, to $I_0^{-1}\beta I_0$, to conclude that $I_1^{-1}I_0^{-1}\beta I_0 I_1$ has power precisely m. Repeating this argument $k-2$ times, we conclude that:

$$I_j^{-1} \cdots I_0^{-1} \beta I_0 \cdots I_j \text{ has power } m \text{ for each } j \in \{0,\cdots,k-1\} .$$

We are now in the situation of Lemma 2.7.7, with $X = I_0 \cdots I_k$ and $I = I_0$. Applying Lemma 2.7.7, we conclude that

$$I_0^{-1}\Delta^m P I_0 = \Delta^m P_1 , \text{ where } P_1 \in C(\Delta^m \hat{P}) .$$

If $k > 2$, Lemma 2.7.7 can be applied a second time, with $\beta = \Delta^m P_1$, $X = I_1 \cdots I_k$, and $I = I_1$. In fact, we can apply Lemma 2.7.7 a total of $k-2$ times, obtaining each time:

$$I_j^{-1}\Delta^m P_j I_j = \Delta^m P_{j+1}$$

where each $P_{j+1} \in C(\Delta^m \hat{P}_j)$, $j = 1,\cdots,k-1$. Thus:

(2-26) $$I_{k-1}^{-1} \cdots I_0^{-1}\Delta^m P I_0 \cdots I_{k-1} = \Delta^m P_k$$

where $P_k \in C(\Delta^m \hat{P}_{k-1})$, $P_{k-1} \in C(\Delta^m \hat{P}_{k-2}),\cdots, P_1 \in C(\Delta^m P)$.

Now observe that, from equations (2-25) and (2-26):

$$I_k^{-1}\Delta^m P_k I_k \text{ has power } m+1 .$$

Since I_k is an initial route, and since $\Delta^m P_k$ is in standard form, Lemma 2.7.4 is applicable, and we conclude that $D(P_k)$ contains a word of the form $I_k\hat{Q}\hat{F}_k$ (if m is even) or $\hat{I}_k\hat{Q}\hat{F}_k$ (if m is odd), where F_k is the final route belonging to the initial route I_k. But then $C(P_k)$ contains a word of the form $\Delta\hat{Q}$.

Finally, we note that since $P_k \in C(\Delta^m\hat{P}_{k-1})$, $P_{k-1} \in C(\Delta^m\hat{P}_{k-2}),\cdots$, $P_1 \in C(\Delta^m P)$, it follows from Lemma 2.7.6 that either $C(P_k) = C(P)$ or $C(\hat{P}_k) = C(P)$. Since $C(P_k)$ contains a word of the form $\Delta\hat{Q}$, and since $C(\hat{P}_k)$ is obtained from $C(P_k)$ by replacing each word in $C(P_k)$ by its reflection, and since the reflection of $\Delta\hat{Q}$ is ΔQ, it follows that $C(P)$ contains either the word ΔQ or the word $\Delta\hat{Q}$. This completes the proof of Theorem 2.7. ||

The proof of Corollary 2.7.1 is an immediate consequence of Theorem 2.7, applied to Garside's algorithm for computing the set $S(\beta)$. Garside's algorithm is described in the text just before the statement of Theorem 2.6. ||

2.4. *The algebraic link problem*

When Artin first suggested braid theory as an approach to the study of knots and links, he conjectured that the chief obstacle in the approach would be the solution to the conjugacy problem in B_n. Since we now have available an algorithmic solution to the conjugacy problem (Theorems 2.6 and 2.7 and Corollary 2.7.1) it is natural to ask whether this might lead to an algorithmic solution to the link problem.

Ideally, one would like to know: Given two oriented polygonal links V_1, V_2, does there exist an algorithm to decide whether V_1 is combinatorially equivalent to V_2? Markov's result (Corollary 2.3.1) shows that this question is indeed equivalent to an algebraic problem; however, the algebraic problem is not simply the conjugacy problem in the group B_n, but rather a problem which concerns the entire sequence of braid groups $B_1, B_2 \cdots$ and which includes as sub-problems the word and conjugacy problems in the individual groups B_n. We define the *algebraic link problem* to be the question: Given two closed braids $\hat{\beta}$ and $\hat{\beta}'$ does there exist an algorithm to decide whether (β', n') can be obtained from (β, n) by a finite sequence of Markov moves $\mathfrak{M}_1$ and $\mathfrak{M}_2$, as defined in Corollary 2.3.1? Since the formulation of this problem splits into two parts, the first question to be settled is whether the application of Markov move $\mathfrak{M}_2$ really causes any serious trouble? That is, it is conceivable that after a sequence of applications of moves $\mathfrak{M}_1$ and $\mathfrak{M}_2$, which increase and decrease string index, but ultimately return a braid to the group B_n from which it originated, one has simply replaced the original braid by a conjugate of itself. We will soon see that in general this is certainly not the case; however, the possibility remains that with suitable restrictions on the class of links under investigation, the link problem might reduce to the conjugacy problem in B_n. The object of this section is to indicate what some of these restrictions must be. At the conclusion of this monograph we give a list of problems encompassing what appear, at the moment, to be the more promising directions for future study. Problems 3 through 12 in particular relate to the results in this chapter.

NOTATIONS AND DEFINITIONS:

As before, $\alpha, \beta, \gamma, \cdots$ will be used to denote elements in B_n, and $\hat{\alpha}, \hat{\beta}, \hat{\gamma}, \cdots$ to denote the oriented link in E^3 determined by the braids $\alpha, \beta, \gamma, \cdots$. If $\gamma = \alpha\beta$ (i.e., if γ is the product of the braids α and β in B_n) we will sometimes use the symbol $(\alpha\beta)\hat{}$ instead of $\hat{\gamma}$.

$\hat{\alpha} \sim \hat{\beta}$ means that α and β determine combinatorially equivalent links in E^3.

$\hat{\alpha} \# \hat{\beta}$ is a symbol which is defined without ambiguity if both $\hat{\alpha}$ and $\hat{\beta}$ are oriented knots, and denotes the composite knot with components $\hat{\alpha}$, $\hat{\beta}$ [see Fox, 1962a]. (If $\hat{\alpha}$ and or $\hat{\beta}$ have multiplicity > 1 the symbol

$\hat{\alpha} \# \hat{\beta}$ can still be defined, but some care is needed to specify the components which are involved. We will not have occasion to use the latter concept.)

A link $\hat{\beta}$ is said to have *braid index* n if it can be represented by a braid word $\beta \in B_n$, but cannot be represented by a braid word in B_{n-1}.

A link $\hat{\beta}$ is said to be a *pure link of multiplicity* n if it is represented by an element β in the pure braid group P_n.

We first ask whether non-conjugate braids with the same string index ever define equivalent links? Our first example shows that this cannot be true unless we restrict our attention to links in B_n which have braid index n:

Example 1. If $\beta \in B_{n-1}$, then $(\beta\sigma_{n-1})\hat{}$ and $(\beta\sigma_{n-1}^{-1})\hat{}$ are equivalent links (by two applications of $\mathfrak{M}_2$), but $\beta\sigma_{n-1}$ is not conjugate to $\beta\sigma_{n-1}^{-1}$ (because they have exponent sums which differ by 2, and exponent sum is an invariant of conjugacy class in B_n.

Difficulties of this type will be ruled out if we restrict our attention to links in B_n which have braid index n. One would expect that it is an extremely difficult problem to determine the braid index of an arbitrary link. In this connection, pure links might be a particularly tractible class because of the following result:

THEOREM 2.8. *Every pure link of multiplicity* n *has braid index* n.

Proof. Let $\beta \in P_n$ be a pure braid representative of the pure link $\hat{\beta}$. Since $\hat{\beta}$ has n components, the group $\pi_1(S^3 - \hat{\beta})$ is generated by at least n elements. If $\hat{\beta}$ could be represented by a braid in B_m for some $m < n$, then (by Theorem 2.2) one could obtain a presentation for $\pi_1(S^3 - \hat{\beta})$ with m generators, which is impossible. Hence $\hat{\beta}$ has braid index n. ||

Continuing our analysis of the algebraic link problem, we consider a second example, which indicates that it may be important to distinguish between braids whose closure is a prime knot or link, and braids whose closure is a composed knot or link:

Example 2 [(Murasugi-Thomas, 1972)]. Let

$$\beta_1 = \sigma_1^3\sigma_2^5\sigma_3^7 \quad \text{and} \quad \beta_2 = \sigma_1^5\sigma_2^3\sigma_3^7 \in B_4 .$$

Then

(i) $\hat{\beta}_1$ and $\hat{\beta}_2$ each have braid index 4
(ii) $\hat{\beta}_1 \sim \hat{\beta}_2$
(iii) β_1 is not conjugate to β_2 in B_4
(iv) $\hat{\beta}_1 = \hat{\alpha} \# \hat{\gamma}$ where α and γ are non-trivial knots.

To prove that $\hat{\beta}_1$ and $\hat{\beta}_2$ satisfy (i)-(iv) above, we first establish:

PROPOSITION 2.9. *Let* $\hat{\beta}$ *be a knot. If* $\beta(\sigma_1,\cdots,\sigma_{n-1}) = \alpha(\sigma_1,\cdots,\sigma_k)\gamma(\sigma_{k+1},\cdots,\sigma_{n-1})$, *for some* $1 \le k \le n-2$, *then* $\hat{\beta} = \hat{\alpha} \# \hat{\gamma}$.

Proof of Proposition 2.9. Examine the schematic diagram of Figure 12. The braid diagrams for α and γ are to be inserted in the boxes labelled α and γ. The dashed curve represents a cross section of a 2-sphere in E^3 which intersects $(\alpha\gamma)\hat{}$ in two points and exhibits $(\alpha\gamma)\hat{}$ as the knot composite of $\hat{\alpha}$ and $\hat{\gamma}$. ||

By the commutativity of composed knots, $\hat{\alpha} \# \hat{\gamma} \sim \hat{\gamma} \# \hat{\alpha}$. Hence it is true that:

$$\begin{aligned}
\hat{\beta}_1 &= (\sigma_1^3\sigma_2^5\sigma_3^7)\hat{} && \text{(word in } B_4) \\
&\sim (\sigma_1^3)\hat{} \# (\sigma_1^5)\hat{} \# (\sigma_1^7)\hat{} && \text{(words in } B_2) \\
&\sim (\sigma_1^5)\hat{} \# (\sigma_1^3)\hat{} \# (\sigma_1^7)\hat{} && \text{(words in } B_2) \\
&\sim (\sigma_1^5\sigma_2^3\sigma_3^7)\hat{} = \hat{\beta}_2 && \text{(word in } B_4) .
\end{aligned}$$

To see that $\hat{\beta}_1$ cannot be represented by a 3-braid, we observe that if this were possible, then the group $\pi_1(S^3-\hat{\beta}_1)$ would have a presentation with three Wirtinger generators (Theorem 2.2). This would imply that

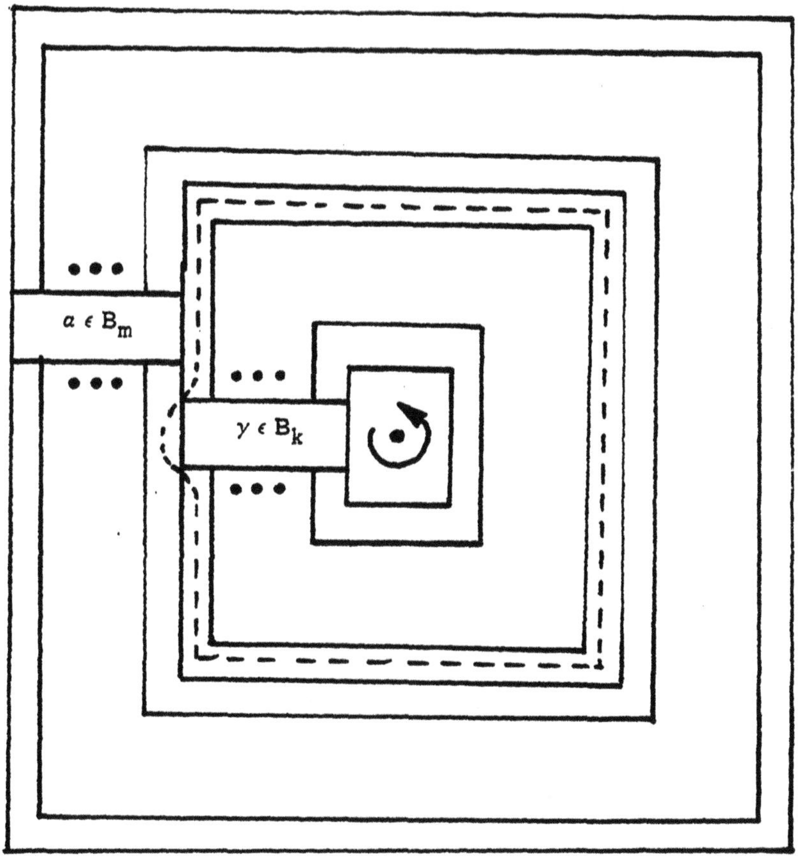

Fig. 12.

the length of the associated chain of elementary ideals is at most 2. But $\hat{\beta}_1$ and $\hat{\beta}_2$ have associated chain of ideals of length 3 because chain length is additive for knot composition, and σ_1^3, σ_1^5 and σ_1^7 each define torus knots, for which chain length is 1 [Ref: R. H. Fox, 1950].

To see that $\hat{\beta}_1$ is not conjugate to $\hat{\beta}_2$ in B_4, observe that a natural homomorphism exists from $B_4 \to B_3$ defined by the mapping $\sigma_1 \to \sigma_1, \sigma_2 \to \sigma_2, \sigma_3 \to \sigma_1$. The images of β_1 and β_2 under this homomorphism have summit representatives $\sigma_1^{10}\sigma_2^5$ and $\sigma_1^{12}\sigma_2^3$ respectively in B_3 (cf. Theorem 2.6). Thus the images of β_1 and β_2 in B_3 are not conjugate, hence β_1 is not conjugate to β_2. This completes Example 2. ||

An essential feature of Example 2 was the fact that knot composition is commutative. The reader who attempts to express the commutativity of composed knots by a sequence of operations of types $\mathfrak{M}_1$ and $\mathfrak{M}_2$ will find that a very complex sequence is required. This suggests that, if we have in mind an algorithmic solution to the algebraic knot problem, it would be wise for us to restrict our attention to *prime knots*. We remark, however, that for the special case of pure links, the question of whether the link is a prime link or a composed link does not appear to lead to the difficulties which are illustrated by Example 2. It would be interesting to learn how prime knots can be characterized algebraically. One would expect that if $\beta \in B_n$, then $\hat{\beta}$ is prime iff β is not conjugate to a "split braid," i.e., a braid word of the form $U(\sigma_1, \cdots, \sigma_k)\, V(\sigma_{k+1}, \cdots, \sigma_{n-1})$.

Example 3. Reversal of knot orientation or of space orientation. Note that a closed braid always has a natural orientation which is induced by a positive sense of rotation about the braid axis, and that if $\beta = \sigma_{s_1}^{n_1} \sigma_{s_2}^{n_2} \cdots \sigma_{s_r}^{n_r} \in B_n$ is a braid word which represents the oriented link L, then $\text{Rev}\,\beta = \sigma_{s_r}^{n_r} \cdots \sigma_{s_2}^{n_2} \sigma_{s_1}^{n_1}$ represents the same link L with reversed orientation. (Similarly, β^{-1} represents the image of L under an orientation-reversing homeomorphism of S^3.) One might expect, however, that a very complex sequence of moves of type $\mathfrak{M}_1$ and $\mathfrak{M}_2$ might be required to take (β, n) to $(\text{Rev}\,\beta, n)$, in the event that L is an invertible link. This is borne out by the example of the invertible knot 6_3 (in Reidemeister's notation), which has the braid representative $\beta = \Delta^{-3}\sigma_1^4\sigma_2^3\sigma_1^2$ in summit form, while $\text{Rev}\,\beta$ has the summit form $\Delta^{-3}\sigma_1^4\sigma_2^2\sigma_1^3$. This suggests that, if we seek a solution to the algebraic link problem, we treat an oriented link, and the same link with its orientation reversed, as special cases which require some sort of special treatment.

An alternative approach, to handle the situation illustrated by Example 3, is to weaken the notion of "combinatorial equivalence of oriented link," and replace it by the notion of "equivalence" [cf. Crowell and Fox, Chapter 1]. In this case, we ask a modified question: if

$\beta, \beta' \in B_n$ are minimal string braid representatives of link types L, L' respectively, can we find appropriate restrictions (e.g., $\hat{\beta}$ is *prime*, etc.) under which L is equivalent to L' if and only if β' is conjugate to *either* β or Rev β or β^{-1} or Rev β^{-1}? (cf. Problem 2, Appendix).

A new class of difficulties occur with links which can be represented by braid words $\beta \in B_{n+1}$ of the form $\beta = U\sigma_n V\sigma_n^{-1}$, where $U, V \in B_n$. A simple picture should convince the reader that if β is as above, and if $\beta' = U\sigma_n^{-1} V\sigma_n$, then $\hat{\beta} \sim \hat{\beta}'$. Yet β and β' need not be conjugate, as is illustrated by Example 4 below:

Example 4. Consider the links $\hat{\beta}$ and $\hat{\beta}'$, which are represented by the pure 4-braids

$$\beta = \sigma_3\sigma_1^2\sigma_2^4\sigma_3^{-1}\sigma_2^2 \quad \text{and} \quad \beta' = \sigma_3^{-1}\sigma_1^2\sigma_2^4\sigma_3\sigma_2^2 .$$

Then $\hat{\beta} \sim \hat{\beta}$ (draw pictures). To see that β is not conjugate to β', consider the images of β and β' in B_3, under the homomorphism defined in Example 2; a calculation just like that in Example 2 proves that β and β' cannot be conjugate in B_4. Details are left to the reader.

The sequence of transformations of types $\mathfrak{M}_1$ and $\mathfrak{M}_2$ which allow one to deform the pair $(\beta, n+1)$ in Example 4 above to $(\beta', n+1)$ can be achieved by first increasing string index, and then lowering it again. The required sequence was discovered by R. Fein (unpublished). It is:

$$\text{(2-27)} \quad (U\sigma_n^{-1}V\sigma_n, n+1) \to (U\sigma_n^{-1}V\sigma_n^{-1}\sigma_n\sigma_n, n+1) \to (U\sigma_n^{-1}V\sigma_n^{-1}\sigma_{n+1}^{-1}\sigma_n\sigma_n, n+2)$$

$$\to (U\sigma_n^{-1}V\sigma_{n+1}\sigma_n^{-1}\sigma_{n+1}^{-1}\sigma_n, n+2)$$

$$\to (U\sigma_n^{-1}\sigma_{n+1}V\sigma_n^{-1}\sigma_{n+1}^{-1}\sigma_n, n+2) \to (U\sigma_{n+1}\sigma_{n+1}^{-1}\sigma_n^{-1}\sigma_{n+1}V\sigma_n^{-1}\sigma_{n+1}^{-1}\sigma_n, n+2)$$

$$\to (U\sigma_{n+1}\sigma_n\sigma_{n+1}^{-1}\sigma_n^{-1}V\sigma_n^{-1}\sigma_{n+1}^{-1}\sigma_n, n+2)$$

$$\to (\sigma_{n+1}U\sigma_n\sigma_{n+1}^{-1}\sigma_n^{-1}V\sigma_n^{-1}\sigma_{n+1}^{-1}\sigma_n, n+2)$$

$$\to (\sigma_{n+1}U\sigma_n\sigma_{n+1}^{-1}\sigma_n^{-1}V\sigma_{n+1}\sigma_n^{-1}\sigma_{n+1}^{-1}, n+2)$$

$$\to (U\sigma_n\sigma_{n+1}^{-1}\sigma_n^{-1}V\sigma_{n+1}\sigma_n^{-1}, n+2)$$

$$\to (U\sigma_n\sigma_{n+1}^{-1}\sigma_n^{-1}\sigma_{n+1}V\sigma_n^{-1}, n+2)$$

$$\to (U\sigma_n\sigma_n\sigma_{n+1}^{-1}\sigma_n^{-1}V\sigma_n^{-1}, n+2)$$

$$\to (U\sigma_n\sigma_n\sigma_n^{-1}V\sigma_n^{-1}, n+1) \to (U\sigma_nV\sigma_n^{-1}, n+1) .$$

We remark that in B_3 a word of the form $U\sigma_2V\sigma_2^{-1}$, where $U, V \in B_2$, has the special form $\beta = \sigma_1^k\sigma_2\sigma_1^m\sigma_2^{-1}$, which can be reduced to $\sigma_1^{k-1}\sigma_2^m\sigma_1$ after several applications of relations (1-1) and (1-2). On the other hand, $\beta' = \sigma_1^k\sigma_2^{-1}\sigma_1^m\sigma_2 = \sigma_1^{k+1}\sigma_2^m\sigma_1^{-1}$. Since β and β' are clearly conjugate in B_3, it follows that difficulties of the type illustrated by Example 4 do not occur for string index < 4.

In view of Example 4, it would be interesting to ask whether there exist classes of braids in B_{n+1} (or in P_{n+1}) which cannot be conjugate to braids of the form $U\sigma_nV\sigma_n^{-1}$, where $U, V \in B_n$? This would seem to be a very difficult problem.

The particular form of Garside's solution to the word and conjugacy problems in B_n (Theorem 2.6) suggests that a class of links which merit further study are links which are the closure of *positive* words in the braid group. We give such links the name *positive links*. It is of interest to note that positive links were studied in 1936 by W. Burau, who called them "gleichsinnig verdrillte Verkettunger." Burau established:

THEOREM 2.10 [Burau, 1936]. *Suppose* $\beta \in B_n$ *is represented by positive word* P, *which involves all of the braid generators* $\sigma_1, \sigma_2, \cdots, \sigma_n$. *Suppose* P *has letter length* m. *Then the Alexander polynomial of* β *has degree* $m-n+1$ *and has leading coefficient* 1.

Proof of Theorem 2.10. Burau's proof depends upon the existence of a matrix representation for B_n which will be introduced in Chapter 3 of the text. We defer the proof until the Appendix to Chapter 3.

CHAPTER 3
MAGNUS REPRESENTATIONS

R. H. Fox's free differential calculus [see R. H. Fox, 1953] can be used to define a number of interesting matrix representations of a free group of finite rank and of various subgroups of the automorphism group of a free group. The first such representation was probably introduced by W. Magnus, 1939, therefore we have given this class of representations the generic name of Magnus representations. Particularly important representations of this type, which will be discussed below in detail, are the Burau representation of Artin's braid group and the Gassner representation of the pure braid group. These representations are of particular interest because the Burau and Gassner matrices of a braid are closely related to the Alexander matrix and reduced Alexander matrix of the corresponding closed braid.

Section 3.1 contains a brief review of the basic facts we will need to know about the free calculus. Propositions 3.1 - 3.4 are all easy consequences of the basic definition of derivation in the group ring of a free group of finite rank. Theorem 3.5 (due to Blanchfield) and Theorem 3.6 (due to Lyndon) are deeper results, which will be needed later in this chapter.

Section 3.2 introduces Magnus representations of a free group, and of the automorphism group of a free group. The results in this section, with a few exceptions, have all been published elsewhere, however our point of view is somewhat more general than that in most of the published literature [e.g., Magnus 1939; Gassner 1961; Bachmuth 1965 and 1966; Chein 1968; Gupta 1968; Remeslennikov and Sokolov, 1970; Gupta and Levin, 1973], and some of our proofs are new.

Section 3.3 is concerned with the Burau and Gassner representation of the braid group. Proposition 3.10, Lemma 3.11.1, Theorem 3.11, and Corollary 3.11.2 how a classical result about knot polynomials may be reproved in a new way, by making use of the theory of braids – many other known results may be similarly re-established, and this suggests the possibility of applying braid theory to certain unsolved problems in link theory (e.g., see problems 15 and 16 in the Appendix).
possibility of applying braid theory to certain unsolved problems in link theory (e.g., see problems 15 and 16 in the Appendix).

The latter half of Section 3.3 (beginning with Proposition 3.12) is concerned with the difficult open question of the faithfulness of the Burau and Gassner representations, and with the implications of faithfulness or non-faithfulness for various questions in link theory. Most of these results are known; some are new, having been obtained by the author in the course of an unsuccessful attempt to establish the faithfulness of these representations.

The appendix to Chapter 3 contains the proof of Theorem 2.10, which was stated in Chapter 2. The proof was deferred until the relevant background material had been developed.

3.1. *Free differential calculus*

The basic facts concerning the free differential calculus [R. H. Fox, 1953] will be reviewed in Propositions 3.1 - 3.4 below. Following this, we will prove two results (Theorems 3.5, 3.6) which are of fundamental importance in the theory of Magnus representations.

Let F_n be a free group of rank n, with free basis $x_1, \cdots, x_n$.[1] Let ϕ be an arbitrary homomorphism acting on F_n, and let F_n^ϕ denote the

---

[1] The abstract group F_n with free basis $x_1, \cdots, x_n$ occurred previously in Corollary 1.8.3 and Theorems 1.9 and 1.10. In those discussions we had in mind a particular geometric realization of F_n, which we will return to later.

image of F_n under ϕ. Let JF_n^ϕ denote the group ring of F_n^ϕ with respect to the ring of integers J. An element of JF_n^ϕ is a sum $\sum a_g g$, $g \in F_n^\phi$, $a_g \in J$. Addition and multiplication in JF_n^ϕ are defined by the rules

$$\sum a_g g + \sum b_g g = \sum (a_g + b_g) g$$

$$\left(\sum a_g g\right) \cdot \left(\sum b_g g\right) = \sum_g \left(\sum_h a_{gh^{-1}} b_h\right) g .$$

We observe that a homomorphism Ψ of the group F_n^ϕ onto a group $F_n^{\psi\phi}$ induces a ring homomorphism from JF_n^ϕ to $JF_n^{\psi\phi}$, which we denote by the same symbol Ψ, where

$$\left(\sum a_g g\right)^\Psi = \sum a_g g^\Psi, \quad g \in F_n^\phi, \; a_g \in J .$$

We will sometimes be interested in the special cases where Ψ is the abelianizer (denoted $\mathfrak{a}$) or the trivializer (denoted $\mathfrak{t}$).

For each $j = 1, \cdots, n$ there is a mapping (well-defined by Proposition 3.2 below):

$$\frac{\partial}{\partial x_j} : JF_n \to JF_n$$

given by

$$\text{(3-1)} \qquad \frac{\partial}{\partial x_j}\left(x_{\mu_1}^{\varepsilon_1} \cdots x_{\mu_r}^{\varepsilon_r}\right) = \sum_{i=1}^{r} \varepsilon_i \delta_{\mu_i, j} x_{\mu_1}^{\varepsilon_1} \cdots x_{\mu_i}^{\frac{1}{2}(\varepsilon_i - 1)},$$

$$\frac{\partial}{\partial x_j}\left(\sum a_g g\right) = \sum a_g \frac{\partial g}{\partial x_j}, \quad g \in F_n, \; a_g \in J ,$$

(where $\varepsilon_i = \pm 1$ and $\delta_{\mu_i, j}$ is the Kronecker δ). Until $\partial/\partial x_j$ has been proved well-defined it will be assumed simply to be defined on words in the symbols $x_1, \cdots, x_n$.

PROPOSITION 3.1:

(i) $\dfrac{\partial x_i}{\partial x_j} = \delta_{i,j}$

$$\text{(ii)} \quad \frac{\partial x_i^{-1}}{\partial x_j} = -\delta_{i,j}\, x_i^{-1}$$

$$\text{(iii)} \quad \frac{\partial (wv)}{\partial x_j} = \left(\frac{\partial w}{\partial x_j}\right)(v)^{\ddagger} + (w)\left(\frac{\partial v}{\partial x_j}\right).$$

Proof. Clear from the definition. ||

PROPOSITION 3.2. *The mapping* $\frac{\partial}{\partial x_j}$ *is well-defined.*

Proof. Suppose $w_1 = wv$ and $w_2 = wx_i^{\varepsilon} x_i^{-\varepsilon} v$ $(\varepsilon = \pm 1)$ as words in F_n. Proposition 3.1 shows immediately that

$$\partial (x_i^{\varepsilon} x_i^{-\varepsilon}) = 0 .$$

Hence a second application of Proposition 3.1 shows that

$$\frac{\partial w_2}{\partial x_j} = \frac{\partial w}{\partial x_j} + w \frac{\partial (x_i^{\varepsilon} x_i^{-\varepsilon})}{\partial x_j} + w x_i^{\varepsilon} x_i^{-\varepsilon} \frac{\partial v}{\partial x_j} = \frac{\partial w}{\partial x_j} + w \frac{\partial v}{\partial x_j} = \frac{\partial w_1}{\partial x_j} .$$

This completes the proof. ||

PROPOSITION 3.3 (Chain Rule). *If* w *and* $v_1, \cdots, v_n$ *are words in* F_n, *with* $w = w(x_1, \cdots, x_n)$, *then*

$$\text{(3-2)} \quad \frac{\partial}{\partial x_j}(w(v_1(x_1,\cdots,x_n),\cdots,v_n(x_1,\cdots,x_n))) = \sum_{k=1}^{n} \left(\frac{\partial w}{\partial v_k}\right)_{v_k = v_k(x_1,\cdots,x_n)} \left(\frac{\partial v_k}{\partial x_j}\right)$$

Proof of Proposition 3.3. Assume that $w = w_1 x_r^{\varepsilon} (\varepsilon = \pm 1)$. Then Proposition 3.1, part (iii), can be used to expand both sides of (3-2) with respect to this factorization of w. By induction on letter length, (3-2) may be assumed for w_1 in place of w. This inductive hypothesis, together with obvious simplifications using Proposition 1, parts (ii) and (iii), then reduces the proof of (3-2) to the proof of

$$(*) \qquad \frac{\partial}{\partial x_j}[v_r^{\varepsilon}(x_1,\cdots,x_n)] = \varepsilon v_r^{\frac{1}{2}(\varepsilon-1)}\frac{\partial v_r}{\partial x_j}\,.$$

The latter can be proved by an easy computation.

PROPOSITION 3.4 ("Fundamental formula" of free calculus): *Let* $v \in JF_n$. *Then*

$$(3\text{-}3) \qquad \sum_{j=1}^{n} \frac{\partial v}{\partial x_j}(x_j-1) = v - v^t\,.$$

Proof.[2] Suppose that $v = x_{\mu_1}^{\epsilon_1} x_{\mu_2}^{\epsilon_2} \cdots x_{\mu_r}^{\epsilon_r}$, $\epsilon_i = \pm 1$, $i = 1,\cdots,r$, and suppose that $\mu_k = j$. We examine the contribution of the letter $x_{\mu_k}^{\epsilon_k}$ to the sum on the LHS in (3-3). Using equation (3-1), we find that:

$$\frac{\partial v}{\partial x_j}(x_j-1) = \cdots + \left(x_{\mu_1}^{\epsilon_1}\cdots x_{\mu_{k-1}}^{\epsilon_{k-1}} x_j^{\epsilon_k}\right) - \left(x_{\mu_1}^{\epsilon_1}\cdots x_{\mu_{k-1}}^{\epsilon_{k-1}}\right) + \cdots$$

independently of whether ϵ_k is $+1$ or -1. It then follows that the sum on the LHS of (3-3) is a telescoping sum, in which all terms cancel except -1 and $+v$. The extension to arbitrary ring elements is immediate. ||

Remark. The fundamental formula (3-3) could have been used to *define* the partial derivatives $\{\frac{\partial v}{\partial x_j};\ j = 1,\cdots,n\}$. This may be seen by observing that $v - v^t$ belongs to the fundamental ideal of JF_n,[3] and that $\{x_1-1,\cdots,x_n-1\}$ is a basis for this ideal, hence $v - v^t$ has a unique expression as a linear combination of the basis elements over JF_n.

---

2 This particular proof of Proposition 3.4 is, to the author's knowledge, new.

3 The fundamental ideal of JF_n is the ideal consisting of all elements of JF_n which are mapped to 0 under the ring homomorphism induced by t.

THEOREM 3.5 [Blanchfield; proof given here is due to Fox, 1953]. *Let* $v \in F_n$. *Then* $(\partial v/\partial x_j)^\phi = 0$, $j = 1,\cdots,n$, *if and only if* $v \in [\mathrm{her}\,\phi, \mathrm{her}\,\phi]$.[4]

Proof. Suppose $v = v_1 v_2 \cdots v_n$, with $v_k = [r_k, s_k] = r_k s_k r_k^{-1} s_k^{-1}$ and $r_k, s_k \in \ker \phi$. Then, by Proposition 3.1

$$\frac{\partial v}{\partial x_j} = \frac{\partial v_1}{\partial x_j} + v_1 \frac{\partial v_2}{\partial x_j} + \cdots + v_1 \cdots v_{m-1} \frac{\partial v_m}{\partial x_j},$$

and

$$\frac{\partial v_k}{\partial x_j} = \frac{\partial r_k}{\partial x_j} + r_k \frac{\partial s_k}{\partial x_j} - r_k s_k r_k^{-1} \frac{\partial s_k}{\partial x_j} - r_k s_k r_k^{-1} s_k^{-1} \frac{\partial r_k}{\partial x_j}, \quad (1 \le k \le m).$$

Since $\phi(r_k) = \phi(s_k) = 1$, it follows that $(\partial v_k/\partial x_j)^\phi = 0$, hence $(\partial v/\partial x_j)^\phi = 0$.

To prove the converse, assume that $v = x_{\mu_1}^{\epsilon_1} \cdots x_{\mu_r}^{\epsilon_r}$ $(\epsilon_i = \pm 1)$ is a freely reduced word in F_n, and that $(\partial v/\partial x_j)^\phi = 0$ for every j. If r = letter length of v is zero then trivially $v \in [\mathrm{her}\,\phi, \mathrm{her}\,\phi]$. We proceed by induction on r.

For each $i\,(1 \le i \le r)$ define: $c_i = \epsilon_i x_{\mu_1}^{\epsilon_1} \cdots x_{\mu_{i-1}}^{\frac{1}{2}(\epsilon_i - 1)}$. Then:

$$0 = (\partial v/\partial x_j)^\phi = \sum_{\mu_i = j} (c_i)^\phi \quad (j = 1,\cdots,n). \tag{3-4}$$

Since each $(c_i)^\phi$ is a monomial in JF_n^ϕ with coefficient $\epsilon_i = \pm 1$, it follows that equation (3-4) can be satisfied only if there is a partitioning of the index set $(1,\cdots,r)$ into pairs $(p_1, q_1),\cdots,(p_s, q_s)$ such that for each integer $t\,(1 \le t \le s)$:

$$\mu_{p_t} = \mu_{q_t} \quad \text{and} \quad (c_{p_t})^\phi = -(c_{q_t})^\phi.$$

---

4 The symbol $[\mathrm{her}\,\phi, \mathrm{her}\,\phi]$ denotes the commutator subgroup of $\mathrm{her}\,\phi$.

The partition of the preceding paragraph may be arranged so that $p_t < q_t$ $(1 \le t \le s)$, $q_1 < q_2 < \cdots < q_s$, and $p_2 = q_1 - 1$. Let $\mu_{p_1} = \mu_{q_1} = \ell$, $\epsilon_{p_1} = -\epsilon_{q_1} = \epsilon$, $\mu_{p_2} = \mu_{q_2} = m$, and $\epsilon_{p_2} = \epsilon_{q_2} = \eta$. Then v is of the form:

$$v = Ax_\ell^\epsilon Bx_m^\eta x_\ell^{-\epsilon} Cx_m^{-\eta} D . \tag{3-5}$$

It follows that:

$$\begin{aligned} c_{p_1} - c_{q_1} &= A(1 - x_\ell Bx_m^\eta x_\ell^{-1}) \quad \text{if} \quad \epsilon = 1 \\ &= -Ax_\ell^{-1}(1 - Bx_m^\eta) \quad \text{if} \quad \epsilon = -1 . \end{aligned} \tag{3-6}$$

Since $(c_{p_1} - c_{q_1})^\phi = 0$, it follows that $Bx_m^\eta \in \ker \Phi$. A similar calculation involving $c_{p_2} - c_{q_2}$ establishes that $x_\ell^{-\epsilon}C \in \ker \phi$. Therefore

$$\begin{aligned} v &= Ax_\ell^\epsilon (Bx_m^\eta x_\ell^{-\epsilon} Cx_m^{-\eta} B^{-1} C^{-1} x_\ell^\epsilon)(x_\ell^{-\epsilon} CBx_m^\eta) x_m^{-\eta} D \\ &= Ax_\ell^\epsilon (x_\ell^{-\epsilon} CBx_m^\eta) x_m^{-\eta} D \bmod [\ker\phi, \ker\phi] \\ &= ACBD \bmod [\ker\phi, \ker\phi] . \end{aligned} \tag{3-7}$$

Since $ACBD$ has shorter letter length than v and since $(\partial(ACBD)/\partial x_j)^\phi = (\partial v/\partial x_j)^\phi = 0$ for each j, it follows that $ACBD \in [\ker\phi, \ker\phi]$, hence $v \in [\ker\phi, \ker\phi]$ and the proof is complete. ||

Let R be the kernel of the group homomorphism $\phi : F_n \to F_n^\phi$. As observed earlier, the group homomorphism from $F_n \to F_n^\phi$ induces a ring homomorphism from $JF_n \to JF_n^\phi$, the latter having as its kernel the ideal $R-1$ generated by all binomials $r-1$, where $r \in R \subseteq F_n$. If $w, v \in JF_n$, and if $w^\phi = v^\phi$, we will write $w \equiv v$ modulo R (rather than modulo $R-1$).

THEOREM 3.6 (Lyndon, 1950; proof here due to Fox, 1953). *Let* $v_1, \cdots, v_n \in JF_n$, *and let* $v - 1 = \sum_{j=1}^{n} v_j(x_j - 1)$. *Then* $v - 1 \equiv 0$ modulo R *iff there exists an element* $r \in R$ *with the properties* $(\partial r/\partial x_j) \equiv v_j$ modulo R *for each* $j = 1, \cdots, n$.

Proof. Suppose there is an element $r \in R$ such that $(\partial r/\partial x_j) \equiv v_j$. By Proposition 3.4, $r-1 = \sum_{j=1}^{n} (\partial r/\partial x_j)(x_j-1)$. Since $r \in R$, it follows that $r-1 \equiv 0$, hence $\sum_{j=1}^{n} (\partial r/\partial x_j)(x_j-1) \equiv 0$, therefore $v-1 = \sum_{j=1}^{n} v_j(x_j-1) \equiv 0$, which proves the "if" part.

Conversely, suppose that

$$(3\text{-}8) \qquad v-1 = \sum_{j=1}^{n} v_j(x_j-1) \equiv 0 .$$

Then it must be possible to write $v-1$ in the form:

$$(3\text{-}9) \qquad v-1 = \sum_{i=1}^{m} \epsilon_i w_i(r_i-1) u_i$$

where $\epsilon_i = \pm 1$; $w_i, u_i, r_i \in F_n$; and $r_i \in R$. The RHS of equation (3-9) is a sum of monomials with coefficients ± 1. Using the relation $-w_i(r_i-1) = w_i(1-r_i) = w_i r_i(r_i^{-1}-1)$ we can rewrite the RHS of (3-9) as a sum of monomials with only positive coefficients, hence we may assume without loss of generality that $\epsilon_i = +1$ for every $i = 1,\cdots,m$ in (3-9). Then:

$$(3\text{-}10) \qquad \frac{\partial v}{\partial x_j} = \sum_{i=1}^{n} w_i \frac{\partial r_i}{\partial x_j} + w_i(r_i-1)\frac{\partial u_i}{\partial x_j} \equiv \sum_{i=1}^{n} w_i \frac{\partial r_i}{\partial x_j}$$

From equation (3-8) we also have

$$\frac{\partial v}{\partial x_j} = v_j .$$

Hence:

$$(3\text{-}11) \qquad \sum_{i=1}^{m} w_i \frac{\partial r_i}{\partial x_j} \equiv v_j \quad (j=1,\cdots,n) .$$

Now define $r = (w_1 r_1 w_1^{-1})(w_2 r_2 w_2^{-1})\cdots(w_m r_m w_m^{-1})$. Since $r_1,\cdots,r_m \in R$, it follows that $r \in R$. Moreover:

(3-12) $$\frac{\partial r}{\partial x_j} \equiv \sum_{i=1}^{m} w_i \frac{\partial r_i}{\partial x_j}$$

Hence, comparing (3-11) and (3-12), we see that $\partial r/\partial x_j \equiv v_j$ $(j = 1, \cdots, n)$, and our proof is complete. ||

3.2. *Magnus representations*

Let $\overline{S}_n$ be a free abelian semi-group with basis $\overline{s}_1, \cdots, \overline{s}_n$, let $\mathcal{R}$ be a ring, and let $A_0(\mathcal{R}, \overline{S}_n)$ be the semi-group ring of $\overline{S}_n$ with respect to the ring $\mathcal{R}$. (elements in $A_0(\mathcal{R}, \overline{S}_n)$ are polynomials in non-negative powers of the commuting indeterminates $\overline{s}_1, \cdots, \overline{s}_n$, with coefficients in $\mathcal{R}$). As before, F_n is the free group with basis $x_1, \cdots, x_n$. We define a mapping τ from F_n to the multiplicative group of 2×2 matrices over $A_0(JF_n, \overline{S}_n)$ by the rule that if $w \in F_n$, then w is mapped to the matrix $[w]$, where

(3-13) $$[w] = \begin{bmatrix} w & \sum_{j=1}^{n} (\partial w/\partial x_j)\overline{s}_j \\ 0 & 1 \end{bmatrix} .$$

In particular, we have:

(3-14) $$[x_i] = \begin{bmatrix} x_i & \overline{s}_i \\ 0 & 1 \end{bmatrix} .$$

If $w, v \in F_n$, then it follows immediately from Proposition 3.1 that:

$$[wv] = [w][v] .$$

Hence the mapping $w \to [w]$ is a representation of F_n, which will be denoted the *Magnus representation of* F_n.

Since w is an entry in the matrix $[w]$, this representation does not appear to be particularly useful as it stands. If, however, we let ϕ be an arbitrary homomorphism acting on F_n, and let $[w]^\phi$ be the matrix whose entries are the images of the entries of w under the ring homomorphism induced by ϕ, i.e.,

$$[w]^{\phi} = \begin{bmatrix} (w)^{\phi} & \sum_{j=1}^{n} (\partial w/\partial x_j)^{\phi}\,\bar{s}_j \\ 0 & 1 \end{bmatrix} \tag{3-15}$$

then the mapping $w \to [w]^{\phi}$ will clearly still be a representation of F_n. Let $[F_n]^{\phi}$ be the image of F_n under this homomorphism. We will call the mapping $\phi : F_n \to [F_n]^{\phi}$ by $w \to [w]^{\phi}$ the *Magnus ϕ-representation of* F_n.

THEOREM 3.7:

(i) *The kernel of the Magnus ϕ-representation is the commutator subgroup of* $\ker \phi$.

(ii) *Let* M *be a* 2×2 *matrix with entries in the ring* $A_0(JF_n^{\phi}, \bar{S}_n)$ *which has the special form:*

$$M = \begin{bmatrix} m & \sum_{j=1}^{n} m_j \bar{s}_j \\ 0 & 1 \end{bmatrix} \quad \textit{where} \quad m \in F_n^{\phi}, \quad m_j \in JF_n^{\phi} .$$

Then $M \in [F_n]^{\phi}$ *iff the elements* $m, m_1, \cdots, m_n$ *satisfy the linear relationship:*

$$\sum_{j=1}^{n} m_j (x_j^{\phi} - 1) = m - 1 . \tag{3-16}$$

(*Remark*: This embedding of $F_n/[\ker\phi, \ker\phi]$ in a matrix ring was first discovered by W. Magnus, who proved Theorem 3.7, part (i) for a special case [Magnus, 1939]. Magnus's representation has proved to be an extremely useful tool in many different contexts, two examples of which are M. Hall's theory of central extension [M. Hall, Theory of Groups, Section 15.5] and D. Cohen's study of laws in a metabelian variety [D. Cohen, 1967]. Theorem 3.7, part (ii) generalizes a result which was first proved by Bachmuth, 1966, for the special case where ϕ is the abelianizing homomorphism. The proof given here is different from Bachmuth's.)

Proof: (i) The matrix $[w]^\phi$ will be the identity matrix iff

(a) $(\partial w/\partial x_j)^\phi = 0 \quad (j = 1, \cdots, n)$

(b) $(w)^\phi = 1$.

Theorem 3.5 tells us that (a) will be satisfied iff $w \in [\ker\phi, \ker\phi]$, and if this is true then condition (b) is also satisfied, hence Theorem 3.7, part (i) is true.

(ii) If the matrix M belongs to $[F_n]^\phi$ then for each $j = 1, \cdots, n$ we have $m = w^\phi$, $m_j = (\partial w/\partial x_j)^\phi$ for some $w \in F_n$. It then follows from Proposition 3.4 that the entries $m, m_1, \cdots, m_n$ satisfy equation (3-16).

Suppose, conversely, that $m, m_1, \cdots, m_n$ satisfy equation (3-16). Let w be any lift of m to F. Consider the matrix $[w^{-1}]^\phi$:

$$[w^{-1}]^\phi = \begin{bmatrix} (w^{-1})^\phi & -(w^{-1})^\phi \sum_{j=1}^{n} (\partial w/\partial x_j)^\phi \bar{s}_j \\ 0 & 1 \end{bmatrix} .$$

Taking products:

$$[w^{-1}]^\phi M = \begin{bmatrix} 1 & (w^{-1})^\phi \sum_{j=1}^{n} (m_j - (\partial w/\partial x_j)^\phi) \bar{s}_j \\ 0 & 1 \end{bmatrix} .$$

Using the fact that the m_j satisfy equation (3-16), and that the entries $(\partial w/\partial x_j)^\phi$ are related by the "fundamental formula of the free calculus," equation (3-3), the entries in the upper right-hand corner of the matrix $[w^{-1}]^\phi M$ are seen to satisfy:

$$\sum_{j=1}^{n} (m_j - (\partial w/\partial x_j)^\phi (x_j^\phi - 1) = (m-1) - (w-1)^\phi = 0 .$$

Therefore, by Theorem 3.6 there is an element $r \in \ker\phi$ which has the property:

$$(\partial r/\partial x_j)^\phi = m_j - (\partial w/\partial x_j)^\phi .$$

Thus the matrix $[w^{-1}]^{\phi}M$ is in $[F_n]^{\phi}$, and since $[w^{-1}]^{\phi}$ is in $[F_n]^{\phi}$, it follows that $M \in [F_n]^{\phi}$. ||

The Magnus ϕ-representation of F_n may be generalized to a representation of F_n by $k \times k$ upper triangular matrices in the following manner. First define higher order derivatives inductively. Writing D_j for $\frac{\partial}{\partial x_j}$, define

$$D_{i_1 i_2 \cdots i_q}(w) = D_{i_q}(D_{i_1 i_2 \cdots i_{q-1}}(w)), \quad w \in JF_n$$

$$D(w) = \sum_{i=1}^{n} D_i(w)\bar{s}_i$$

$$D^{q+1}(w) = D(D^q(w)), \quad w \in A_0(JF_n, \bar{S}_n) .$$

An easy induction establishes that

$$\text{(3-17)} \qquad \begin{aligned} D^{q+1}(w) &= \sum_{1 \leq i_j \leq n} D_{i_1 i_2 \cdots i_q}(w)\bar{s}_{i_1}\bar{s}_{i_2}\cdots\bar{s}_{i_q} \\ D^q(uv) &= \sum_{p=1}^{q-1} (D^p(u))(D^{q-p}(v))^t + uD^q(v) . \end{aligned}$$

Again, let ϕ be a homomorphism acting on F_n, and let $(D^q(w))^{\phi}$ be the image of $D^q(w)$ under the ring homomorphism induced by ϕ. Then we have

THEOREM 3.8 [Enright, 1968]. *Let* $w \in F_n$, *and define*

$$\{w\}^{\phi} = \begin{bmatrix} w^{\phi} & (D(w))^{\phi} & (D^2(w))^{\phi} & (D^3(w))^{\phi} & \cdots & (D^{k-1}(w))^{\phi} \\ 0 & 1 & (D(w))^t & (D^2(w))^t & \cdots & (D^{k-2}(w))^t \\ 0 & 0 & 1 & (D(w))^t & \cdots & (D^{k-3}(w))^t \\ \cdot & \cdot & \cdot & \cdot & & \cdot \\ \cdot & \cdot & \cdot & \cdot & & \cdot \\ \cdot & \cdot & \cdot & \cdot & & \cdot \\ 0 & 0 & 0 & 0 & \cdots & 1 \end{bmatrix}$$

Then the mapping $w \to \{w\}^{\phi}$ *defines a representation of* F_n *in the ring of* $k \times k$ *matrices over* $A_0(JF_n^{\phi}, \bar{S}_n)$, *for every integer* $k \geq 2$ *and for every homomorphism* ϕ *acting on* F_n.

Proof. It is only necessary to check that if $w, u \in F_n$, then $\{wu\}^{\phi} = \{w\}^{\phi}\{u\}^{\phi}$. This is a direct consequence of the product rule in equation (3-17). ||

COROLLARY 3.8.1. *Let* x_i *be a basis element of* F_n, *and define* $\{x_i\}^t$ *to be the* $k \times k$ *matrix*

$$\{x_i\}^t = \begin{bmatrix} 1 & s_i & 0 & \cdots & 0 \\ 0 & 1 & s_i & \cdots & 0 \\ \cdot & \cdot & \cdot & & \\ \cdot & \cdot & \cdot & & \\ \cdot & \cdot & \cdot & & \\ 0 & 0 & 0 & \cdots & s_i \\ 0 & 0 & 0 & \cdots & 1 \end{bmatrix} \left\{ \begin{array}{l} k \text{ rows and columns} \\ k \geq 2 \\ 1 \leq i \leq n \end{array} \right.$$

Then the mapping $x_i \to \{x_i\}^t$ *is a faithful matrix representation of* F_n *modulo the* k^{th} *group of the lower central series of* F_n, *over the ring* $A_0(J, \bar{S}_n)$, *for every integer* $k \geq 2$.

Proof. The fact that $x_i \to \{x_i\}^t$ defines a representation of F_n follows directly from the Theorem 3.8, setting $\phi = t$. The proof that the kernel of the representation is the kth group of the lower central series follows from [Theorem 4.6 of Fox, 1953], where it is established that a nasc for the vanishing of $(D^j(w))^t$ for each $j = 0, \cdots, k-1$ is that w belongs to the kth group of the lower central series of F_n. ||

Remark. The representation in Corollary 3.8.1 may be regarded as the matrix equivalent of the "Taylor Series" representation of F_n [see Fox, 1952] or of the "Magnus embedding" of F_n in $A_0(J, \bar{S}_n)$ [see Magnus, Karass and Solitar, 1966, p. 308]. The problem of identifying the kernel of the *generalized Magnus ϕ-representation*, as defined in Theorem 3.8, if $\phi \neq t$ and if $k \geq 3$ appears to be a messy problem, with an unpleasant answer. The special cases $k = 3$ and $k = 4$ were treated in Enright's thesis, but the solutions he found are too unappealing to repeat. Other special cases have been treated by [Gupta, 1969][5] and again by [Gupta and Levin, 1973].[5] The clue to a deeper understanding of the generalized representations would appear to be in the proper generalization of Theorem 3.5 to the case where the condition $(\frac{\partial}{\partial x_j})^\phi = 0$, $j = 1,\cdots,n$ is replaced by $(D^j(v))^\phi = 0$ for each $j = 1,\cdots,k-1$. (See Problem 13, Appendix.)

Our purpose in introducing Magnus representations has been not for the study of quotient groups of F_n, but rather for the study of subgroups of the automorphism group of F_n, in particular Artin's braid group and pure braid group. We show next that matrix representations of certain subgroups of Aut F_n can be defined with the aid of free calculus.

As before, let ϕ be an arbitrary homomorphism acting on the free group F_n of rank n. Let A_ϕ be any group of (right) automorphisms of F_n which satisfy the condition

$$x\phi = x\alpha\phi \tag{3-18}$$

for each $x \in F_n$, $\alpha \in A_\phi$. (For example, if ϕ is the abelianizer, we could choose A_ϕ to be the subgroup of those automorphisms of F_n which map each element into a conjugate of itself; indeed, any subgroup of Aut F_n which induces the identity automorphism on F_n/F'_n would do.) If $\alpha \in A_\phi$, we define $\|\alpha\|$ to be the $n \times n$ matrix

$$\|\alpha\|^\phi = \left[\left(\frac{\partial(x_i\alpha)}{\partial x_j}\right)^\phi\right] \tag{3-19}$$

[5] The representations considered by Gupta and by Gupta and Levin differ in detail from the representation in Theorem 3.8, but may be interpreted in a similar fashion in terms of the free calculus. The identification of the kernels of all such representations would be greatly facilitated by a solution of Problem 13 in the Appendix of this monograph.

with entries in JF_n^{ϕ}. The mapping $\tau : \alpha \to \|\alpha\|^{\phi}$ is called a Magnus representation of A_{ϕ}. This representation is closely related to the representation of F_n defined in equation (3-16): explicitly, the entry in the ith row and j^{th} column of $\|\alpha\|^{\phi}$ is the coefficient of $\bar{s}_j$ in the entry in the upper right-hand corner of $[x_i\alpha]^{\phi}$.

THEOREM 3.9. *The mapping* $\tau : \alpha \to \|\alpha\|^{\phi}$ *defines a group homomorphism from* A_{ϕ} *into the multiplicative group of* $n \times n$ *matrices over* JF_n^{ϕ}.

Proof. Suppose $\alpha \in A_{\phi}$ is the identity automorphism. Then:

$$\text{(3-20)} \qquad \left(\frac{\partial(x_i\alpha)}{\partial x_j}\right)^{\phi} = \delta_{ij} \; .$$

Hence $\|\alpha\|^{\phi}$ will be the identity matrix.

Suppose now that $\alpha, \beta \in A_{\phi}$ are arbitrary, with $x_i\alpha = w_i(x_1, \cdots, x_n)$ and $x_i\beta = v_i(x_1, \cdots, x_n)$. Then by Proposition 3.3:

$$\text{(3-21)} \qquad \frac{\partial}{\partial x_j}(x_i\alpha\beta) = \frac{\partial}{\partial x_j}(w_i(v_1(x_1, \cdots, x_n), \cdots, v_n(x_1, \cdots, x_n)))$$

$$= \sum_{k=1}^{n} \left(\frac{\partial w_i}{\partial x_k}\right)_{x_r = v_r(x_1, \cdots, x_n)} \left(\frac{\partial v_k}{\partial x_j}\right) .$$

But $x\phi = x\beta\phi$ for each $x \in F_n$; thus, in particular:

$$\text{(3-22)} \qquad \left(\left(\frac{\partial w_i}{\partial x_k}\right)_{x_r = v_r = x_r\beta}\right)^{\phi} = \left(\frac{\partial w_i}{\partial x_k}\right)^{\phi} = \left(\frac{\partial(x_i\alpha)}{\partial x_k}\right)^{\phi} .$$

It follows immediately from (3-21) and (3-22) that

$$\left(\frac{\partial}{\partial x_j}(x_i\alpha\beta)\right)^{\phi} = \sum_{k=1}^{n} \left(\frac{\partial w_i}{\partial x_k}\right)^{\phi} \left(\frac{\partial v_k}{\partial x_j}\right)^{\phi} = \sum_{k=1}^{n} \left(\frac{\partial(x_i\alpha)}{\partial x_k}\right)^{\phi} \left(\frac{\partial(x_k\beta)}{\partial x_j}\right)^{\phi}$$

or that $\|\alpha\beta\|^{\phi} = \|\alpha\|^{\phi}\|\beta\|^{\phi}$ as desired. ||

We now give several examples of Magnus representations of subgroups of Aut F_n.

Example 1. Let ϕ be the trivial homomorphism, $t: F_n \to 1$, and let A_t be the full group Aut F_n. The resulting Magnus representation is surprisingly non-trivial. If $\alpha \in$ Aut F_n, and $x_i\alpha = w_i(x_1, \cdots, x_n)$, then it follows from the basic formula (3-1) that $(\partial w_i/\partial x_j)^t$ is the exponent sum of the letter x_j in the word w_i. Thus our representation maps each element of Aut F_n into an $n \times n$ matrix of integers, where the entry in the ith row, j^{th} column is the exponent sum of x_j in w_i. Since the matrix is invertible, it must have determinant ± 1. Thus the image of F_n is a subgroup of the unimodular group. Clearly, each $n \times n$ matrix with integral entries and determinant ± 1 can be associated with an automorphism of the free abelian group, F_n/F'_n. Now, it was proved by [Nielsen, 1924, pp. 169-209] that every automorphism of a free abelian group of rank n is induced by an automorphism of a free group of rank n, hence the Magnus representation of Aut F_n is a faithful representation of the full group Aut (F_n/F'_n) by the full group of $n \times n$ unimodular matrices.

Example 2 [cf. Bachmuth, 1965]. Let ϕ be the abelianizer, $\mathfrak{a}$, and let $A_{\mathfrak{a}}$ be the largest possible subgroup of Aut F_n which satisfies equation (3-18), with $\phi = \mathfrak{a}$. The elements in this group are known as "I-A automorphisms" of F_n, where the acronym "I-A" denotes the fact that all automorphisms in the group induce the identity automorphism in the abelianized group F_n/F'_n. It follows immediately from Proposition 3.4 that the rows of a matrix in the Magnus representation of $A_{\mathfrak{a}}$ must satisfy the "fundamental formula"; it is also an immediate consequence of Theorem 3.5 that the matrices in the group $A_{\mathfrak{a}}$ are in 1-1 correspondence with the automorphisms of F_n/F''_n which are induced by I-A automorphisms of F_n. Even more, one can use Theorem 3.6 to prove that *every* $n \times n$ invertible matrix over $JF_n^{\mathfrak{a}}$ whose rows satisfy the fundamental formula corresponds to a I-A automorphism of F_n/F''_n. (These results were first established by other methods in Bachmuth, 1965.) We denote the group of matrices with this property by $\hat{A}_{\mathfrak{a}}$.

Suppose one now enlarges the concept of a I-A automorphism, and defines the full group of I-A automorphisms of F_n/F''_n to be the group of all automorphisms of F_n/F''_n which induce the identity automorphism of F_n/F'_n. The I-A automorphisms of F_n/F''_n are in 1-1 correspondence with the elements in the matrix group $\hat{A}_\alpha$, which includes the matrix group A_α. One is prompted to ask whether A_α coincides with $\hat{A}_\alpha$, i.e., is every I-A automorphism of F_n/F''_n induced by a (I-A) automorphism of F_n? The surprising answer (see Chein, 1968) is no, i.e., A_α is a *proper* subgroup of $\hat{A}_\alpha$. (Of course, every matrix in A_α is induced by some *endomorphism* of F_n; indeed, the proof of Theorem 3.6 contains instruction for constructing such an endomorphism.)

Example 3. The *Burau representation* of the braid group B_n [Burau, 1936]. By Corollary 1.8.3, B_n has a faithful representation as a group of right automorphisms of the free group F_n of rank n. This allows us to regard B_n as a subgroup of Aut F_n. Explicitly, we choose generators $\sigma_1,\cdots,\sigma_{n-1}$ of B_n with the automorphisms $(\sigma_1)\xi,\cdots,(\sigma_{n-1})\xi$ given by equation (1-14). As in Chapter 1, we will write σ_i instead of $(\sigma_i)\xi$.

Let $Z = \langle t \rangle$ be an infinite cyclic group, and let $\psi : F_n \to Z$ be defined by $x_i\psi = t$, $1 \le i \le n$. The condition of equation (3-18), i.e., $x_j\sigma_i\psi = x_j\psi$, is satisfied for all pairs (i, j) of interest, hence there is a Magnus representation of B_n over JF_n^ψ. The matrix $\|\sigma_i\|^\psi$ corresponding to the generator σ_i of B_n is given by:

$$\text{(3-23)} \qquad \|\sigma_i\|^\psi = \left[\begin{array}{c|cc|c} I_{i-1} & 0 & 0 & 0 \\ \hline 0 & 1-t & t & 0 \\ 0 & 1 & 0 & 0 \\ \hline 0 & 0 & 0 & I_{n-i-1} \end{array}\right] \begin{array}{l} \\ i^{th} \text{ row} \\ (i+1)^{st} \text{ row} \\ \\ \end{array}$$

where I_k denotes the $k \times k$ identity matrix. This representation of B_n is named for its discoverer, [W. Burau, 1936], and is therefore known as the Burau representation of B_n. We will use the symbol $\|B_n\|^{\psi}$ to denote the image of B_n in the ring of matrices over JF_n^{ψ}, the symbols $\|\beta\|^{\psi}, \|\gamma\|^{\psi}, \cdots$ to denote elements of $\|B_n\|^{\psi}$, and τ_B to denote the homomorphism from $B_n \to \|B_n\|^{\psi}$ defined by $\tau_B(\beta) = \|\beta\|^{\psi}$.

Example 4. The *Gassner representation* of the pure braid group.

Recall that the pure braid group P_n is defined to be the subgroup of B_n consisting of those elements of B_n which belong to the kernel of the natural homomorphism γ from $B_n \to \Sigma_n$ (cf. Chapter 1). Generators of P_n are related to those of B_n by equations (1-11). Clearly these equations could be used to define a representation of P_n by matrices, simply by computing the images of the pure braid generators A_{rs} under the mapping $\sigma_i \to \|\sigma_i\|^{\psi}$ defined by (3-23).

We find, however, that a slightly "better" representation of P_n can be obtained if we attempt to apply Theorem 3.9 directly to P_n. The group P_n has a faithful representation as a subgroup of Aut F_n, with the action of the pure braid generators A_{rs} on the generators $x_1, \cdots, x_n$ of F_n being given by equation (1-15) of Chapter 1.[6] Let Z_n be a free abelian group of rank n with free basis $t_1, \cdots, t_n$ and let $\alpha F_n \to Z_n$ be defined by $x_i\alpha = t_i$, $1 \leq i \leq n$. Since the pure braid generators map each generator x_i of F_n into a conjugate of itself, the condition of equation (3-18), i.e., $x_i A_{rs} \alpha = x_i \alpha$ will be satisfied for every $1 \leq i \leq n$, $1 \leq r < s \leq n$ if we choose the homomorphism ϕ of equation (3-18) to be α, hence the mapping $A_{rs} \to ((A_{rs}))^{\alpha}$, defined by:

$$((A_{rs}))^{\alpha} = \left[\left(\frac{\partial(x_i A_{rs})}{\partial x_j}\right)^{\alpha}\right] \tag{3-24}$$

$$\begin{aligned}
\left(\frac{\partial(x_i A_{rs})}{\partial x_j}\right)^{\alpha} &= \delta_{ij} \quad (\text{if } s < i \text{ or } i < r) \\
&= (1-t_i)\delta_{ir} + t_r\delta_{ij} \quad (\text{if } s = i) \\
&= (1-t_i)(\delta_{ij} + t_i\delta_{sj}) + t_i t_s \delta_{ij} \quad (\text{if } r = i) \\
&= (1-t_i)(1-t_s)\delta_{rj} - (1-t_r)\delta_{sj} + \delta_{ij} \quad (\text{if } r < i < s)
\end{aligned}$$

6 We use the symbol A_{rs} instead of $(A_{rs})\xi$ in this section.

will define a Magnus representation of P_n. This representation was first discovered by B. J. Gassner, 1961, hence we refer to it as the Gassner representation. Since it will be of particular interest to us in which follows, we adopt the special symbol $((P_n))^a$ to denote the matrix group corresponding to P_n, the symbols $((\beta))^a$, $((\gamma))^a, \cdots$ for elements of $((P_n))^a$, and τ_G for the homomorphism from P_n to $((P_n))^a$.

We will use the symbols $\tau_{G,n}$, $\tau_{B,n}$ instead of τ_G, τ_B when we wish to stress string index n. We will also use the symbols τ (or τ_n) when we wish to refer simultaneously to *either* τ_G or τ_B (or to $\tau_{G,n}$ or $\tau_{B,n}$).

We remark that the matrix group $\|B_n\|^\psi$ contains a subgroup of index n! which is a homomorphic image of $((P_n))^a$. The homomorphism from $((P_n))^a$ into $\|B_n\|^\psi$ is defined by replacing the n indeterminates $t_1, \cdots, t_n$ in $((P_n))^a$ by the single indeterminate t. For this reason it is more likely that $\ker \tau_{G,n} = 1$ than that $\ker \tau_{B,n} = 1$.

3.3. *The Burau and Gassner representations*

Our object now is to study the Burau representation of Artin's braid group $B_n = \pi_1 B_{0,n}$ and the Gassner representation of the pure braid group $P_n = \pi_1 F_{0,n}$, Examples 3 and 4 of Section 3.2. The early results in the section (Proposition 3.10 and Theorem 3.11) explore the relationship between the matrices in these groups and the Alexander matrices of knots and links. We will then turn our attention to the open question about whether these representations are faithful.

It will be assumed that the reader is familiar with the concept of the Alexander matrix of a link. (For a brief introduction to this subject, see R. H. Fox, 1962a, Sections 4 and 5.)

PROPOSITION 3.10. *If* $\beta \in B_n$ *is such that* $\hat{\beta}$ *is a knot, then* $\|\beta\|^\psi - \text{Id}$ *is an Alexander matrix for* $\hat{\beta}$. *If* $\beta \in P_n$, *then* $((\beta))^a - \text{Id}$ *is an Alexander matrix for the pure link* $\hat{\beta}$.[7]

---

[7] If $\beta \in B_n$, and if $\hat{\beta}$ is a link of $\mu \geq 1$ components, then the matrix $[\|\beta\|^\psi - I]$ is a reduced Alexander matrix for $\hat{\beta}$, that is, it is the matrix obtained from the Alexander matrix by replacing the indeterminates $t_1, \cdots, t_\mu$ in that matrix by a single indeterminate, t.

Proof. Let the action of β on the free group F_n be given by equation (1-22). Then, by Theorem 2.2, $\pi_1(S^3-\hat{\beta})$ admits the presentation (2-2). It then follows that the matrix:

$$\left[\frac{\partial(A_i x_{\mu_i} A_i^{-1} x_i^{-1})}{\partial x_j}\right]$$

is a Jacobian matrix for $\hat{\beta}$. Proposition 3.10 follows immediately. ||

The following lemma will be of use:

LEMMA 3.11.1. *The Burau and Gassner representations are reducible to* $(n-1)\times(n-1)$ *representations.*

Proof. In $F_n = \langle x_1,\cdots,x_n\rangle$ introduce the free generating set $\langle g_1,\cdots,g_n\rangle$ with $g_i = x_1x_2\cdots x_i$. Recall the homomorphism $\psi: F_n \to Z = \langle t\rangle$ from F_n onto the infinite cyclic group Z defined in Example 3: $x_i\psi = t$ and $g_i\psi = t^i$. The generators $\sigma_1,\cdots,\sigma_{n-1}$ of B_n act on $g_1,\cdots,g_n$ as follows:

$$\text{(3-25)}\qquad \sigma_i : \begin{cases} g_k \to g_k & (\text{if } (k\neq i)) \\ g_i \to g_{i+1}g_i^{-1}g_{i-1} & (\text{if } (i\neq 1)) \\ g_1 \to g_2g_1^{-1} & (\text{if } (i=1)) . \end{cases}$$

The corresponding Magnus representation of B_n, with respect to the new basis $g_1,\cdots,g_n$, will be $\beta \to \left[\left(\frac{\partial(g_i\beta)}{\partial g_j}\right)^{\psi}\right]$. In particular, the generators $\sigma_1,\cdots,\sigma_{n-1}$ will be mapped as follows:

$$\text{(3-26)}\qquad \sigma_1 \to \begin{bmatrix} -t & 1 & & 0 \\ 0 & 1 & & 0 \\ & & \ddots & \\ 0 & 0 & & 1 \end{bmatrix} \qquad \sigma_r \to \begin{bmatrix} 1 & & & & & \\ & \ddots & & & & \\ & & 1 & 0 & 0 & \\ & & t & -t & 1 & \\ & & 0 & 0 & 1 & \\ & & & & & \ddots \ \ 1 \end{bmatrix} \begin{matrix} \\ \\ \\ \leftarrow r^{th} \text{ row} \\ (1<r<n) \\ \\ \end{matrix}$$

These matrices can also be obtained by replacing the Burau matrices $\|\beta\|^{\psi}$ by their conjugates $C\|\beta\|^{\psi}C^{-1}$, where:

$$(3\text{-}27)\qquad C = [(\partial g_i/\partial x_j)^{\psi}] = \begin{bmatrix} 1 & 0 & 0 & \cdots & 0 \\ 1 & t & 0 & \cdots & 0 \\ 1 & t & t^2 & \cdots & 0 \\ \cdot & \cdot & \cdot & & \\ \cdot & \cdot & \cdot & & \\ \cdot & \cdot & \cdot & & \\ 1 & t & t^2 & \cdots & t^{n-1} \end{bmatrix}$$

The last row in the mapping defined by (3-26) is always $(0\ 0 \cdots 1)$. Hence the last row and column may be deleted to obtain an $(n-1)\times(n-1)$ representation of B_n. A similar proof applies to P_n. $\|$ We will denote the image of an element $\beta \in B_n$ or P_n in this $(n-1)\times(n-1)$ representation by the symbols $\|\beta\|_r^{\psi}$ and $((\beta))_r^{a}$.

THEOREM 3.11 (a modification of a result due to Burau, 1936). *If* $\beta \in B_n$, *and if* $\tilde{\nabla}_\beta(t)$ *is the reduced Alexander polynomial of* $\hat{\beta}$ *(see footnote 7, p. 120), then*

$$(3\text{-}28)\qquad (1+t+\cdots+t^{n-1})\tilde{\nabla}_\beta(t) = \det[\|\beta\|_r^{\psi} - \mathrm{Id}]\ .$$

If $\beta \in P_n$, *and if* $\Delta_\beta(t_1,\cdots,t_n)$ *is the Alexander polynomial of* $\hat{\beta}$, *then:*

$$(3\text{-}29)\qquad (1+t_1+t_1t_2+\cdots+t_1t_2\cdots t_{n-1})\Delta_\beta(t_1,\cdots,t_n) = \det[((\beta))_r^{a} - \mathrm{Id}]\,.$$

Proof. By Proposition 3.10, the matrix $A_\beta(t) = [C\|\beta\|^{\psi}C^{-1} - \mathrm{Id}]$ is a reduced Alexander matrix for β. Since the last row of $C\|\beta\|^{\psi}C^{-1}$ is $(0,\cdots,0,1)$, the last row of $A_\beta(t)$ will be $(0,\cdots,0,0)$, hence $\tilde{\Delta}_\beta(t)$ is the greatest common divisor of the determinants $D_k(t)$ of the $(n-1)\times(n-1)$ matrices obtained by deleting the kth row and k^{th} column from $A_\beta(t)$. The determinant $D_n(t)$ is, of course, $\det[\|\beta\|_r^{\psi} - \mathrm{Id}]$. To prove Theorem 3.11, we must see how $D_1(t),\cdots,D_{n-1}(t)$ are related to $D_n(t)$.

The entry in the ith row and j^{th} column of $C\|\beta\|^{\psi}C^{-1}$ is the image under ψ of the free derivative $\partial(g_i\beta)/\partial g_j$, hence the entries in the ith row of $C\|\beta\|^{\psi}C^{-1}$ must satisfy the fundamental formula of free calculus (Proposition 3.4), hence:

$$\sum_{j=1}^{n} (\partial(g_i\beta)/\partial g_j)^{\psi}(t^j-1) = t^i-1 \qquad 1 \leq i \leq n-1 . \tag{3-30}$$

Hence, if we let $A_\beta(t) = \|m_{ij}\|$, then:

$$\sum_{j=1}^{n} m_{ij}(t^j-1) = 0 . \tag{3-31}$$

Keeping equation (3-31) in mind, we now investigate how the determinants $D_k(t)$, $(1 \leq k \leq n)$, are related.

$$D_k(t) = \begin{vmatrix} m_{11} & \cdots m_{1,k-1} m_{1,k+1} \cdots & m_{1,n} \\ \cdot & & \cdot \\ \cdot & & \cdot \\ \cdot & & \cdot \\ m_{n-1,1} & \cdots m_{1,k-1} m_{1,k+1} \cdots & m_{n-1,n} \end{vmatrix}$$

$$= (-1)^{n-k+1} \begin{vmatrix} m_{1,1} & \cdots & m_{1,k-1} & m_{1,n} & m_{1,k+1} & \cdots & m_{1,n-1} \\ \cdot & & & & & & \cdot \\ \cdot & & & & & & \cdot \\ \cdot & & & & & & \cdot \\ m_{n-1,1} & \cdots & m_{n-1,k-1} & m_{n-1,n} & m_{n-1,k+1} & \cdots & m_{n-1,n-1} \end{vmatrix}$$

Observe that $D_k(t)$ differs from $D_n(t)$ only in the kth column. Applying (3-31), and adding (t^j-1) times the jth column of $D_k(t)$ to the kth column of $D_k(t)$ for every $j \neq k$, we obtain

$$(t^n-1)D_k(t) = (-1)^{n-k}(t^k-1)D_n(t)$$

or

$$(1+t+\cdots+t^{n-1})D_k(t) = (-1)^{n-k}(1+t+\cdots+t^{k-1})D_n(t) .$$

If we now set $\tilde{\tilde{\nabla}}_\beta(t) = (1+t+\cdots+t^{n-1})^{-1}D_n(t)$ then this implies

$$D_k(t) = (-1)^{n-k}(1+t+\cdots+t^{k-1})\tilde{\tilde{\nabla}}_\beta(t) \cdot \quad 1 \leq k \leq n .$$

Hence $\tilde{\tilde{\nabla}}_\beta(t)$ is the g.c.d. of the polynomials $D_k(t)$, $1 \le k \le n$, that is, $\tilde{\Delta}_\beta(t) = \tilde{\tilde{\Delta}}_\beta(t)$, and

$$D_n(t) = (1+t+\cdots+t^{n-1})\tilde{\nabla}_\beta(t)\,. \tag{3-32}$$

Since $D_n(t) = \det\,[\|\beta\|_r^\psi - \mathrm{Id}]$, this proves part (i) of Theorem 3.11. The proof of part (ii) is similar. ||

COROLLARY 3.11.1. *Let* $\beta \in P_n$. *Then* $\Delta_\beta(t_1,\cdots,t_n) = 0$ *iff* $\lambda = 1$ *is an eigenvalue of the* $(n-1)\times(n-1)$ *Gassner matrix* $((\beta))_r^a$.

Proof. This follows directly from equation (3-29).

A result which is analogous to Proposition 3.10 and Corollary 3.11.1 applies to arbitrary links, if we look at the appropriate "in between" representations of those subgroups of B_n consisting of braids whose associated permutations admit a decomposition into a product of μ disjoint cycles for some $\mu = 2,\cdots,n-1$, by matrices whose entries are polynomials in μ indeterminates.

Lemma 3.11.1, Theorem 3.11 and Corollary 3.11.1 have some rather curious implications which are not at all well-understood. The mapping defined by equation (3-26) gives an $n\times n$ representation of B_n which is conjugate to the Burau representation. In this latter representation the last row is a unit vector. Thus 1 is an eigenvalue of every Burau matrix. Reducing the representation by deleting this last row and the corresponding column, we obtain an $(n-1)\times(n-1)$ representation which has characteristic polynomial

$$p(x) = |\,\|\beta\|_r^\psi - xI\,|\,,\quad \beta \in B_n\,.$$

According to Theorem 3.11, the reduced Alexander polynomial of the knot $\hat{\beta}$ is, apart from the multiplicative factor $(1+t+\cdots+t^{n-1})$, simply $p(1)$, that is, the remainder on dividing $p(x)$ by $x-1$. Hence the Alexander polynomial of $\hat{\beta}$ is in some sense a measure of how far "1" departs from being an eigenvalue of $\|\beta\|_r^\psi$.

One further observation is in order. Since $p(x)$ is an invariant of the conjugacy class of β in B_n, one might hope that $p(x)$ completely determines conjugacy class in B_n. This seems to be false:

COROLLARY 3.11.2. *The characteristic polynomial of the reduced Burau matrix* $\|\beta\|_r^\psi$ *does not determine the conjugacy class of* β *in* B_n *uniquely.*

Proof. We claim that the 3-braids $\beta_1 = \sigma_1^{-1}\sigma_2^2\sigma_1^{-2}\sigma_2$ and $\beta_2 = \sigma_2\sigma_1^{-2}\sigma_2^2\sigma_1^{-1}$ are not conjugate in B_3, yet they have identical characteristic polynomials $p_i(x)$, $i = 1,2$. To establish that β_1 and β_2 are not conjugate in B_3, we compute their summit forms (cf. Section 2.3), which may be verified to be $\Delta^{-3}\sigma_1^4\sigma_2^3\sigma_1^2$ and $\Delta^{-3}\sigma_1^4\sigma_2^2\sigma_1^3$ respectively. One may verify that the characteristic polynomials $p_1(x)$ and $p_2(x)$ coincide by direct computation, or by the following observations: The knots $\hat{\beta}_1$ and $\hat{\beta}_2$ are equivalent (both represent the invertible knot 6_3), hence they have the same Alexander polynomials $\Delta_1(t) = \Delta_2(t)$. For a 3-braid, we have

$$p_i(x) = x^2 - x\,(\text{trace}\,\|\beta_i\|_r^\psi) + \det\|\beta_i\|_r^\psi\,.$$

By Theorem 3.11 the Alexander polynomial is given by

$$(\Delta_i(t))(1+t+t^2) = 1 - \text{trace}\,\|\beta_i\|_r^\psi + \det\|\beta_i\|_r^\psi = p(1)\,.$$

But $\det\|\beta_i\|_r^\psi$ is just $(-t)$(exponent sum β_i), hence $\det\|\beta_1\|_r^\psi = \det\|\beta_2\|_r^\psi$, and since $\Delta_1(t) = \Delta_2(t)$ it follows that $\text{trace}\,\|\beta_1\|_r^\psi = \text{trace}\,\|\beta_2\|_r^\psi$, hence $p_1(x) = p_2(x)$. ||

We remark that many of the known properties of Alexander matrices of links could be established with the aid of the theory of braids. An example is given below, in Corollary 3.11.2.

COROLLARY 3.11.2 [A new proof will be given of a classical result]. *Let* $\beta \in B_n$, *and suppose that* $\hat{\beta}$ *is a proper knot. Then* $\Delta_\beta(1) = 1$.

Proof. Since $\hat{\beta}$ is a knot, the permutation associated with the braid $\beta \in B_n$ must be an n-cycle; by replacing β with a conjugate of itself, if necessary, we may assume without loss of generality that this permutation is in fact the n-cycle $(1\ 2 \cdots n)$.

Observe that if the defining homomorphism ϕ in the Magnus representation of $B_n \subset \operatorname{Aut} F_n$ is chosen to be the trivializer, then we obtain a representation of the symmetric group Σ_n on n letters by integer matrices. This representation factors through the Burau representation of B_n, hence on setting $t = 1$ in the Burau representation, or in the reduced Burau representation, we obtain a representation of Σ_n, which is easily proved to be faithful.

Using equation (3-26), we may compute the reduced Burau matrix for the braid $\sigma_1\sigma_2 \cdots \sigma_{n-1}$, which induces the permutation $(1\ 2 \cdots n)$, to be:

$$\text{(3-33)} \qquad \|\sigma_1\sigma_2\cdots\sigma_{n-1}\|_r^{\psi} = \begin{bmatrix} 0 & 0 & 0 & \cdots & 0 & -t \\ t & 0 & 0 & \cdots & 0 & -t \\ 0 & t & 0 & \cdots & 0 & -t \\ \cdot & \cdot & \cdot & & \cdot & \cdot \\ \cdot & \cdot & \cdot & & \cdot & \cdot \\ 0 & 0 & 0 & \cdots & t & -t \end{bmatrix}$$

where the matrix in (3-33) has $(n-1)$ rows and columns. It then follows that if $\beta \in B_n$ is any braid which induces this same permutation, then from equation (3-28),

$$\text{(3-34)} \qquad n\Delta_\beta(1) = \begin{vmatrix} -1 & 0 & 0 & \cdots & 0 & -1 \\ 1 & -1 & 0 & \cdots & 0 & -1 \\ 0 & 1 & -1 & \cdots & 0 & -1 \\ \cdot & \cdot & \cdot & & \cdot & \cdot \\ \cdot & \cdot & \cdot & & \cdot & \cdot \\ 0 & 0 & 0 & \cdots & 1 & -2 \end{vmatrix} .$$

Expanding by the first column of the matrix in (3-34) and using the symbol Δ_n to denote the $(n-1)\times(n-1)$ determinant in (3-34), we find:

$$\Delta_n = -\Delta_{n-1} + (-1)^{n-1} .$$

Since $\Delta_2 = -2$, therefore $\Delta_n = (-1)^{n-1}n$, and $\Delta_\beta(1) = \pm 1.$ ||

A similar technique may be used to give new proofs of the theorem about Alexander polynomials of links which were established by [Torres, 1953], by making use of the ideas in Corollary 3.14.1 and Theorem 3.18 below. For lack of space we will not dwell on this aspect of the theory, although it is tempting! This might, however, be a fruitful approach to certain unsolved problems, e.g., the characterization of link polynomials (cf. Problem 2, Appendix).

We now turn our attention to the question of whether the Burau or Gassner representation is faithful. This is an open problem, and the results which follow provide only a partial answer. If either representation is non-faithful, the class of links which are the closure of non-trivial elements in $\ker \tau$ would have a number of very bizarre and interesting properties; on the other hand, if $\ker \tau_G$ or $\ker \tau_B = 1$, these representations might be useful in attacking some of the problems discussed in Chapter 2. We remark that a computer search by the author has failed to turn up any non-trivial elements in $\ker \tau_B$. However, the practical difficulties involved in this search make us hesitate to conjecture strongly that none exist.

We begin with several easy observations.

PROPOSITION 3.12. *If* $\ker \tau_B = 1$, *then* $\ker \tau_G = 1$.

Proof. Set $t_1 = t_2 = \cdots = t_n = t$ in the Gassner representation to obtain the Burau representation. ||

PROPOSITION 3.13. $\ker \tau_B \subset P_n$.

Proof. If one sets the indeterminate t equal to 1 in the matrix group $\|B_n\|^\psi$, then the Burau matrices go over to a faithful representation of the group Σ_n of permutations on n letters, with the matrix $\|\sigma_i\|^\psi$ going over to a matrix which represents the transposition $(i, i+1)$. Since the canonical homomorphism $\gamma : B_n \to \Sigma_n$ maps $\sigma_i \to (i, i+1)$, it follows that a braid which is not in $\ker \gamma = P_n$ cannot have the identity Burau matrix. ||

THEOREM 3.14. *$\ker \tau_B \subset B'_n \cap P_n$ and $\ker \tau_G \subset P'_n$.*[8]

Proof. Examining the matrix $\|\sigma_i\|^\psi$ defined by (3-23), we see that $\det \|\sigma_i\|^\psi = -t$. It then follows that if $\beta \in \ker \tau_B$, and if $\beta = \sigma_{\mu_1}^{\epsilon_1}\sigma_{\mu_2}^{\epsilon_2}\cdots\sigma_{\mu_r}^{\epsilon_r}$, then $\det \|\beta\|^\psi = (-t)^{\epsilon_1+\epsilon_2+\cdots+\epsilon_r}$. Hence $\|\beta\|_\psi = \mathrm{Id}$ implies $\epsilon_1+\epsilon_2+\cdots+\epsilon_r = 0$. Since the commutator quotient group of B_n is infinite cyclic,[9] it then follows that $\beta \in B'_n$.

Similarly, a calculation based on (3-24) shows that $\det((A_{rs}))^a = t_r t_s$ for every pair (r, s) of interest. Hence $\beta \in \ker \tau_G$ only if the exponent sum of each pure braid generator A_{rs} in the braid word β is zero. This implies that β is in the commutator subgroup P'_n of the pure braid group P_n. ||

As an immediate consequence of Theorem 3.14 we obtain a geometric restriction on elements in $\ker \tau_G$:

COROLLARY 3.14.1. *If $\beta \in \ker \tau_G$, then every linking number of $\hat{\beta}$ is zero. If $\beta \in \ker \tau_B$, the sum of its linking numbers is zero.*

Proof. A simple picture (cf. Figure 4) based on the definition of A_{rs} in equation (1-11) shows that each appearance of $A_{rs}^{\pm 1}$ in the braid word β

---

8 The symbols B'_n, P'_n denote the commutator subgroups of B_n, P_n.

9 This may be verified by adding to relations (1-1) and (1-2) the relations $[\sigma_i, \sigma_j] = 1$, $1 \leq i, j \leq n-1$.

contributes ± 1 to the linking number of the rth and s^{th} components of β. By Theorem 3.14, $\beta \in \ker \tau_G$ only if the exponent sum of A_{rs} in β is zero, and this implies that every linking number of $\hat{\beta}$ is zero. For elements in $\ker \tau_B$ we have the weaker result that the exponent sum of β as a word the generators $\{A_{rs};\ 1 \leq r \leq s \leq n\}$ of P_n is zero, which implies that the sum of the linking numbers is zero. ||

For $n = 2$ the braid group is infinite cyclic, and a brief calculation shows that no powers of the matrices which generate either $\|B_2\|^{\psi}$ or $((P_2))^{\alpha}$ are the identity, hence both representations are faithful for $n = 2$. Our next result shows that both representations are also faithful for $n = 3$. The difficulties begin with $n = 4$ (see Theorem 3.19).

THEOREM 3.15 [Magnus and Peluso, 1969]. *If* $n = 3$, *both the Burau and Gassner representations are faithful.*

Proof. We will prove that the Burau representation of B_3 is faithful; using Proposition 3.12 we then obtain immediately that the Gassner representation of P_3 is also faithful.

The group $\|B_3\|_r^{\psi}$ is generated by the matrices:

$$\|\sigma_1\|_r^{\psi} = \begin{pmatrix} -t & 1 \\ 0 & 1 \end{pmatrix} \text{ and } \|\sigma_2\|_r^{\psi} = \begin{pmatrix} 1 & 0 \\ t & -t \end{pmatrix}. \tag{3-35}$$

If we set $t = -1$, these matrices become

$$s_1 = \begin{pmatrix} 1 & 1 \\ 0 & 1 \end{pmatrix} \text{ and } s_2 = \begin{pmatrix} 1 & 0 \\ -1 & 1 \end{pmatrix}. \tag{3-36}$$

By [Coxeter-Moser, 1964, p. 85] s_1 and s_2 generate the homogeneous modular group M_2 (the group of all 2×2-matrices with integral entries and determinant $+1$), which has defining relations (i) $s_1 s_2 s_1 = s_2 s_1 s_2$ and (ii) $(s_1 s_2 s_1)^4 = 1$. Therefore, B_3 maps homomorphically onto M_2. As a result, the only possible relators in $\|B_3\|_r^{\psi}$ arise from (1-2) or a product of conjugates of $\|\sigma_1 \sigma_2 \sigma_1\|^{4\lambda}$, λ an integer. A calculation shows that $\|\sigma_1\sigma_2\sigma_1\|^4 = \begin{pmatrix} t^6 & 0 \\ 0 & t^6 \end{pmatrix}$ is in the center of $\|B_3\|_r^\psi$ and has infinite order. Therefore,

$$\|B_3\|_r^\psi = <\|\sigma_1\|^\psi, \|\sigma_2\|^\psi; \|\sigma_1\|^\psi \|\sigma_2\|^\psi \|\sigma_1\|^\psi = \|\sigma_2\|^\psi \|\sigma_1\|^\psi \|\sigma_2\|^\psi>$$

hence $\|B_3\|_r^\psi$ is isomorphic with B_3. ||

Summarizing our results so far, we have learned that if there exists an element $\beta \in P_n$ such that $\beta \neq 1$, $\beta \in \ker \tau_G$, then:

(i) $\hat{\beta}$ is a link of multiplicity $\mu \geq 4$ (because $\beta \in P_n$, and by Theorem 3.15, $n \geq 4$.

(ii) All the linking numbers of $\hat{\beta}$ are zero (by Corollary 3.14.1).

(iii) All the components of $\hat{\beta}$ are unknotted (because $\hat{\beta}$ is a pure link).

(iv) $\pi_1(S^3 - \hat{\beta})$ has rank $\mu = n$ (by Theorem 2.2).

Our next result (Theorem 3.16 and Corollary 3.16.1) will imply the following additional property:

(v) The Alexander matrix of $\hat{\beta}$ which corresponds to the presentation of Theorem 2.2 will be a matrix of zeros. In particular, the Alexander polynomial of $\hat{\beta}$ will be zero.

We note that while there are many known examples of links with Alexander polynomial zero [see Smythe 1965, also Cochran 1970; also cf. Corollary 3.11.1 above], none of the known examples satisfies conditions (i)-(v). It is easy to prove that the only known *class* of links with zero Alexander polynomial, the so-called "homology boundary links" studied by Smythe, cannot satisfy both conditions (i) and (iv) above.[10] Yet there does not

---

[10] A link is a homology boundary link iff its group maps homomorphically onto the free group of rank μ. The existence of a homology boundary link satisfying (i) and (iv) would then imply the existence of a non-trivial homomorphism from $F_\mu \to F_\mu$, and since free groups are Hopfian [Magnus, Karass and Solitar, 1966] this is impossible.

seem to be any geometrical basis on which we can rule out the existence of non-trivial elements in $\ker \tau_G$. Geometric conditions which are necessary and sufficient conditions for the Alexander polynomial to be zero are given in [Cochran and Crowell, 1970]; however, these seem to be very difficult to apply in any concrete situation.

To establish property (v), let us recall that we studied the question of the faithfulness of the Magnus representation of F_n in Section 3.2 (Theorem 3.7). A result similar to Theorem 3.7 can, of course, be stated for the Magnus representations of subgroups of $\mathrm{Aut}\, F_n$, and we now give this latter result, specialized to the particular cases of the Burau and Gassner representations. In view of Theorem 3.11, there is no loss in generality in restricting our attention to elements in the pure braid group. We have:

THEOREM 3.16. *Let $\beta \in P_n$, and suppose that the action of β on the free group F_n is given by equations (1-22) and (1-23). It will be assumed that the words A_i in equation (1-22) are freely reduced, except for the possibility that an arbitrary factor x_i^m, $m = 0, \pm 1, \pm 2, \cdots$ may be added to the end of each such word, so that the word $A_i x_i A_i^{-1}$ need not be freely reduced. Then:*

$$\beta \in \ker \tau_B \text{ if and only if } A_i \in [\ker\psi, \ker\psi],\ 1 \leq i \leq n\,.$$
$$\beta \in \ker \tau_G \text{ if and only if } A_i \in [\ker a, \ker a],\ 1 \leq i \leq n\,.$$

Proof. Consider the Burau representation first. Recall that $\ker \tau_B \subset P_n$. Hence the $(i,j)^{\text{th}}$ entry in the matrix $\|\beta\|^{\psi}$ is:

$$\left(\frac{\partial(A_i x_i A_i^{-1})}{\partial x_j}\right)^{\psi} = \delta_{ij} t^e + (1-t)\left(\frac{\partial A_i}{\partial x_j}\right)^{\psi} \tag{3-37}$$

where e is the exponent sum of the letters x_k in the word A_i. Examining the off-diagonal entries first, one finds that since the ring JF_n^{ψ} contains no zero divisors, it follows that the off-diagonal entries will be zero iff $\left(\frac{\partial A_i}{\partial x_j}\right)^{\psi} = 0$ for every $1 \le i, j < n$, $i \ne j$. For the diagonal entry, observe that it can always be arranged that $e = 0$ by adding or subtracting additional powers of x_i at the interface between A_i and x_i in $A_i x_i A_i^{-1}$. Hence the diagonal entry will be 1 if and only if $\left(\frac{\partial A_i}{\partial x_i}\right)^{\psi} = 0$. It then follows from Theorem 3.5 that $\|\beta\|^{\psi} = \mathrm{Id}$ if and only if $A_i \epsilon [\ker\psi, \ker\psi]$ for each $i = 1, \cdots, n$.

For the Gassner representation, an identical argument shows that the off-diagonal entries will be zero if and only if $\left(\frac{\partial A_i}{\partial x_j}\right)^{\alpha} = 0$ for each $i \ne j$. The condition that the diagonal entries be 1 is:

$$A_i^{a} + (1-t_i)\left(\frac{\partial A_i}{\partial x_i}\right)^{\alpha} = 1 . \tag{3-38}$$

Setting $t_i = 1$ in equation (3-38) shows the exponent sum on each letter x_j in the word A_i (except possibly $j = i$) must be equal to zero. However, as before, one can always add additional powers of x_i at the end of A_i in order to ensure that the exponent sum of x_i is zero in A_i; hence the diagonal entry is 1 if and only if:

$$A_i \epsilon \ker(\alpha) \text{ and } (1-t_i)\left(\frac{\partial A_i}{\partial x_i}\right)^{\alpha} = 0 . \tag{3-39}$$

This implies that $\left(\frac{\partial A_i}{\partial x_i}\right)^{\alpha} = 0$. It then follows from Theorem 3.5 that $A_i \epsilon [\ker \alpha, \ker \alpha]$ for each $i = 1, \cdots, n$. ||

COROLLARY 3.16.1. *If $\beta \epsilon \ker \tau_{G,n}$, $\beta \ne 1$, then $\pi_1(S^3 - \hat{\beta})$ has a presentation with the property that the Alexander matrix of the presentation is an $(n-1) \times n$ matrix of zeros. In particular, the Alexander polynomial of $\hat{\beta}$ is zero. If $\beta \epsilon \ker \tau_{B,n}$, one obtains the weaker result that the reduced Alexander matrix of the presentation is an $(n-1) \times n$ matrix of zeros.*

Proof. If the action of β on F_n is as in equation (1-22), then by Theorem 2.2 the group $\pi_1(S^3-\beta)$ admits the presentation: (2-3). If $\beta \in \ker \tau_G$ (respectively τ_B), then by Theorem 3.14: $A_i \in [\ker \alpha, \ker \alpha]$ (respectively $[\ker\psi, \ker\psi]$). Corollary 3.14.1 then follows immediately by an application of Theorem 3.5. ||

One would expect, at first glance, that Theorem 3.16 contains instructions for constructing non-trivial braid automorphisms which have the identity Burau and/or Gassner matrices. The difficulty in following these instructions lies in the fact that a pure braid automorphism must not only map each generator into a conjugate of itself, but must also preserve the product $x_1x_2\cdots x_n$ (cf. equation (1-23)), that is, we require that:

$$A_1x_1A_1^{-1}A_2x_2A_2^{-1}\cdots A_nx_nA_n^{-1} = x_1x_2\cdots x_n \tag{1-24}$$

where equality means "freely equal" in F_n. Thus our problem reduces to the question: Do there exist elements $A_1,\cdots,A_n \in F_n$, with $A_i \in [\ker\psi, \ker\psi]$ $(i=1,\cdots,n)$ (or $A_i \in [\ker \alpha, \ker \alpha]$ $i=1,\cdots,n$), which satisfy condition (1-24)? This seems to be a very difficult question; our approach is indirect, i.e., we try to place further restrictions on the elements in $\ker \tau_G$ and $\ker \tau_B$.

It will be shown next (Theorem 3.17) that the kernel of the homomorphism $\tau_{G,n}$ is included in the second commutator subgroup of a certain free subgroup, R_{n-1}, of rank $n-1$ in the pure braid group P_n. The subgroup R_{n-1} is contained in a subgroup, R, of P_n which is defined below.

Let F_n (as before) denote the free group of rank n on which P_n acts. Let $\hat{F}_{n-1}$ denote a free group of rank $n-1$ with free basis $\hat{x}_1,\cdots,\hat{x}_{n-1}$, and let $\chi: F_n \to \hat{F}_{n-1}$ be defined by $\chi(x_i) = \hat{x}_i$, $1 \le i \le n-1$, and $\chi(x_1x_2\cdots x_n) = 1$. If we denote $\chi(x_n) = \hat{x}_n$, we then have $\hat{x}_n^{-1} = \hat{x}_1\hat{x}_2\cdots\hat{x}_{n-1}$. Since $\ker \chi$ is mapped into itself by every braid automorphism, the action of P_n on F_n induces a group P_n^* of automorphisms of $\hat{F}_{n-1}$. Let χ_* denote the homomorphism from $P_n \to P_n^*$ which is induced by χ. The subgroup R is defined to be the group of all braid automorphisms in P_n whose images in P_n^* are inner automorphisms of $\hat{F}_{n-1}$.

To determine R explicitly, and to understand its relationship to $\ker \tau_{G,n}$, we first study the relationship between P_n and P_n^*, and the structure of P_n^*. Let $\hat{P}_{n-1} \subset \mathrm{Aut}\, \hat{F}_{n-1}$ denote the group of pure braid automorphisms of $\hat{F}_{n-1}$, i.e., the subgroup of $\mathrm{Aut}\, \hat{F}_{n-1}$ which maps each $\hat{x}_i$ into a conjugate of itself and preserves the product $\hat{x}_1\hat{x}_2\cdots\hat{x}_{n-1}$ (cf. Theorem 1.9). Let $Z(\hat{P}_{n-1})$ denote the center of $\hat{P}_{n-1}$, which is an infinite cyclic group generated by $\hat{\sigma} = (\hat{\sigma}_1\hat{\sigma}_2\cdots\hat{\sigma}_{n-2})^{n-1}$. Let P_{n-1} denote the subgroup of P_n consisting of all pure braid automorphisms of the subgroup F_{n-1} of F_n generated by $x_1,\cdots,x_{n-1}$. Finally, let R_{n-1} denote the subgroup of P_n generated by $C_1,\cdots,C_{n-1}$, where:

$$C_k = B_{1k}B_{k+1.n}^{-1} \qquad 1 \le k \le n-1 \tag{3-40}$$

$$B_{jk} = (A_{j,j+1})(A_{j,j+2}A_{j+1,j+2})\cdots(A_{jk}A_{j+1,k}\cdots A_{k-1,k}) \text{ if } 1 \le j \le k \le n$$

$$B_{11} = B_{nn} = 1 .$$

(See equations (1-11) and (1-15) for the definition of A_{ij}.)

LEMMA 3.17.1 [Magnus, 1934].

(i) $P_n^* = \hat{P}_{n-1} \cdot \mathrm{Inn}\, \hat{F}_{n-1}$

(ii) $\hat{P}_{n-1} \cap \mathrm{Inn}\, \hat{F}_{n-1} = Z(\hat{P}_{n-1})$

(iii) $\hat{P}_{n-1} = \chi_*(P_{n-1})$ *and* $\chi_*|P_{n-1}$ *is* $1-1$

(iv) $\mathrm{Inn}\, \hat{F}_{n-1} = \chi_*(R_{n-1})$ *and* $\chi_*|R_{n-1}$ *is* $1-1$.

Proof. To obtain the decomposition of P_n^* as $\hat{P}_{n-1} \cdot \mathrm{Inn}\, \hat{F}_{n-1}$, choose any $\hat{\beta} \in P_n^*$. Since every element of P_n^* is induced by an automorphism in P_n, we can lift $\hat{\beta}$ to $\beta \in P_n$. Suppose that β maps $x_i \to A_i x_i A_i^{-1}$, $1 \le i \le n$, and $x_1x_2\cdots x_n \to x_1x_2\cdots x_n$. Let $\hat{A}_i = \chi(A_i)$. Define $\hat{\beta}_2$ to be the inner automorphism of $\hat{F}_{n-1}$ that maps each $\hat{x}_i \to \hat{A}_n\hat{x}_i\hat{A}_n^{-1}$. Define

$\hat{\beta}_1$ to be the automorphism of $\hat{F}_{n-1}$ that maps each $\hat{x}_i \to \hat{A}_n^{-1}\hat{A}_i\hat{x}_i\hat{A}_i^{-1}\hat{A}_n$. Then $\hat{\beta} = \hat{\beta}_1\hat{\beta}_2$, where $\hat{\beta}_2 \in \text{Inn}\,\hat{F}_{n-1}$. To see that $\hat{\beta}_1$ is a *braid* automorphism, observe that $\hat{\beta}$ maps $\hat{x}_1\hat{x}_2\cdots\hat{x}_{n-1} = \hat{x}_n^{-1}$ onto $\hat{A}_n\hat{x}_n^{-1}\hat{A}_n^{-1} = \hat{A}_n\hat{x}_1\cdots\hat{x}_{n-1}\hat{A}_n^{-1}$, hence $\hat{\beta}_1$ maps $\hat{x}_1\hat{x}_2\cdots\hat{x}_{n-1} \to \hat{x}_1\hat{x}_2\cdots\hat{x}_{n-1}$, therefore $\hat{\beta}_1 \in \hat{P}_{n-1}$.

To see that $\hat{P}_{n-1} \cap \text{Inn}\,\hat{F}_{n-1} = Z(P_{n-1})$, note that every braid automorphism must satisfy conditions (1-23), hence the only possible inner automorphisms of $\hat{F}_{n-1}$ are conjugation by powers of $\hat{x}_1\hat{x}_2\cdots\hat{x}_{n-1}$. A calculation based on equation (1-14) shows that the powers of $\hat{\sigma} = (\hat{\sigma}_1\hat{\sigma}_2\cdots\hat{\sigma}_{n-2})^{n-1}$ have the following effect:

$$\text{(3-41)} \qquad \hat{\sigma} : \hat{x}_i \to \hat{x}_1\hat{x}_2\cdots\hat{x}_{n-1}\,\hat{x}_i\hat{x}_{n-1}^{-1}\cdots\hat{x}_2^{-1}\hat{x}_1^{-1}, \quad 1 \le i \le n-1\,.$$

Note that $\hat{\beta}_1$ is only unique up to an arbitrary power of $\hat{\sigma}$, because if we defined $\hat{\beta}'_1 = \hat{\beta}_1\hat{\sigma}^{\ell}$, $\hat{\beta}'_2 = \hat{\sigma}^{-\ell}\hat{\beta}_2$ we would still have $\hat{\beta} = \hat{\beta}'_1\hat{\beta}'_2$, $\hat{\beta}'_1 \in \hat{P}_{n-1}$, $\hat{\beta}'_2 \in \text{Inn}\,\hat{F}_{n-1}$.

To see that $\hat{P}_{n-1} = \chi_*(P_{n-1})$ and that $\chi_*|P_{n-1}$ is 1–1, note that P_{n-1} is generated by the elements $\{A_{ij};\ 1 \le i < j \le n-1\}$, where the action of the A_{ij} on F_n is given by equation (1-15). Observing that the image of x_i $(1 \le i \le n-1)$ under A_{ij} $(1 \le i < j \le n-1)$ does not involve the letter x_n, it follows immediately that $\chi_*|P_{n-1}$ is a 1-1 mapping of P_{n-1} into $\hat{P}_{n-1}$.

To see that every element of $\text{Inn}\,\hat{F}_{n-1}$ is induced by a braid automorphism in R_{n-1}, consider the images of the generators $C_1,\cdots,C_{n-1}$ of R_{n-1} under χ_*. A brief calculation based on the action given in (1-15) shows that the image $\chi_*(C_k)$ of C_k under χ_* is the inner automorphism of $\hat{F}_{n-1}$ corresponding to conjugation by $\hat{x}_1\cdots\hat{x}_k$, hence $\hat{C}_1,\cdots,\hat{C}_{n-1}$ generate the full group $\text{Inn}\,\hat{F}_{n-1}$, hence $\text{Inn}\,\hat{F}_{n-1} = \chi_*(R_{n-1})$.

To see that $\chi_*|R_{n-1}$ is 1–1, note that R_{n-1} is generated by $n-1$ elements and maps homomorphically onto the free group $\text{Inn}\,\hat{F}_{n-1}$ of rank $n-1$. If R_{n-1} were not free, this would imply the existence of a proper homomorphism from the free group of rank $n-1$, factoring through R_{n-1}, onto $\operatorname{Inn} \hat{F}_{n-1}$, which is impossible because free groups are Hopfian [Magnus, Karass and Solitar, 1966, Theorem 2.13]. ||

LEMMA 3.17.2.

(i) $\ker \chi_* = Z(P_n)$.

(ii) $R = R_{n-1} \times Z(P_n)$.

By definition, $R = \chi_*^{-1}(\operatorname{Inn} \hat{F}_{n-1})$. In the proof of Lemma 3.17.1 we showed that $\operatorname{Inn} \hat{F}_{n-1} = \chi_*(R_{n-1})$, hence $\chi_*^{-1}(\operatorname{Inn} \hat{F}_{n-1}) = (R_{n-1})(\ker \chi_*)$, hence the second statement will be true if the first is true.

To see that $\ker \chi_* = Z(P_n)$, the reader may use the known facts that $\hat{P}_{n-1}$ is defined by the braid relations, that $\operatorname{Inn} \hat{F}_{n-1}$ is free, and that the action of $\hat{P}_{n-1}$ on $\operatorname{Inn} \hat{F}_{n-1}$ can be computed by the representation of P_n^* in $\operatorname{Aut} \hat{F}_{n-1}$. This calculation shows that defining relations for P_n^* are the braid relations and $(\hat{\sigma}_1 \hat{\sigma}_2 \cdots \hat{\sigma}_{n-1})^n = 1$. ||

The relationship between the subgroup R of P_n and the kernel of the Gassner representation can now be established: Let R''_{n-1} denote the second commutator subgroup of R_{n-1}. Let $\ker r_{G,n-1}$ denote the intersection of $\ker r_{G,n}$ with the subgroup P_{n-1} of braid automorphisms in P_n which leave the generator x_n fixed.

THEOREM 3.17 [cf. Lipschitz, 1961]. $\ker r_{G,n} \subset (\ker r_{G,n-1})(R''_{n-1})$.

Proof. Choose an arbitrary element $\beta \in \ker r_{G,n}$, where β maps x_i onto $A_i x_i A_i^{-1}$, $1 \le i \le n$. We wish to show that $\beta = \beta_1 \beta_2$, where $\beta_1 \in \ker r_{G,n-1}$ and $\beta_2 \in R''_{n-1}$.

Since $\beta \in \ker r_{G,n}$, it follows from Theorem 3.16 that $A_i \in F''_n = [\ker \alpha, \ker \alpha]$, whence

$$A_i = \prod_{m=1}^{m_i} [\eta_{im1}, \eta_{im2}] \quad \eta_{im\ell} \in F'_n,\ \ell = 1,2, \quad 1 \le i \le n\,.$$

Since $\eta_{im\ell} \in F'_n$, the exponent sum of each letter x_j $(j=1,\cdots,n)$ in $\eta_{im\ell}$ must be zero. From the definition of χ, it then follows that the image $\hat{\eta}_{im\ell}$ of $\eta_{im\ell}$ under χ will contain each letter $\hat{x}_j$ $(j=1,\cdots,n-1)$ with exponent sum zero, hence $\hat{\eta}_{im\ell} \in \hat{F}'_{n-1}$ for every i, m, ℓ of interest, hence $\hat{A}_i \in \hat{F}''_{n-1}$ for each $1 \leq i \leq n$. Therefore the induced automorphism $\hat{\beta}$ of $\hat{F}_{n-1}$ will map each $\hat{x}_i$ onto $\hat{A}_i\hat{x}_i\hat{A}_i^{-1}$, where $\hat{A}_i \in F''_{n-1}$, $1 \leq i \leq n-1$.

Following the procedure of Lemma 3.17.1, we define $\hat{\beta}_2$ to be the inner automorphism of $\hat{F}_{n-1}$ which maps each $\hat{x}_i$ to $\hat{A}_n\hat{x}_i\hat{A}_n^{-1}$, and $\hat{\beta}_1$ to be the braid automorphism of $\hat{F}_{n-1}$ which maps each $\hat{x}_i$ to $\hat{A}_n^{-1}\hat{A}_i\hat{x}_i\hat{A}_i^{-1}\hat{A}_n$. Since $\hat{A}_n^{-1}\hat{A}_i \in F''_{n-1}$ for every $i = 1,\cdots,n-1$, the automorphism $\hat{\beta}_1$ must be in $\ker \hat{r}_{G,n-1} = \chi(\ker r_{G,n-1})$. (See Lemma 3.17.1, (iii).) Since $\text{Inn}\, \hat{F}_{n-1}$ is naturally isomorphic to $\hat{F}_{n-1}$, and since $\hat{A}_n \in \hat{F}''_{n-1}$, therefore $\hat{\beta}_1 \in (\text{Inn}\, \hat{F}_{n-1})''$. Thus

$$\text{(3-42)} \qquad \hat{\beta} = \hat{\beta}_1\hat{\beta}_2 \qquad \hat{\beta}_1 \in \ker \hat{r}_{G,n-1}, \; \hat{\beta}_2 \in (\text{Inn}\, \hat{F}_{n-1})''.$$

Hence

$$\text{(3-43)} \qquad \beta = \beta_1\beta_2 \qquad \beta_1 \in \chi_*^{-1}(\ker \hat{r}_{G,n-1}), \; \beta_2 \in \chi_*^{-1}(\text{Inn}\, \hat{F}_{n-1})''.$$

But $\chi_*|P_{n-1}$ is 1-1 (see Lemma 3.17.1, (iii)), therefore $\beta_1 \in \ker r_{G,n-1}$. Since $\chi_*^{-1}(\text{Inn}\, \hat{F}_{n-1})$ is the direct product of $Z(P_n)$ and R_{n-1} (Lemma 3.17.2), therefore $\chi_*^{-1}(\text{Inn}\, \hat{F}_{n-1})$ must coincide with R''_{n-1}, because any element in the center of P_n cancels when commutators are formed. This completes the proof of Theorem 3.17. ||

In particular, if $n = 4$, we have that $\ker r_{G,4} \subset R''_3$, because $\ker r_{G,3} = 1$ (Theorem 3.15).

To improve upon the result in Theorem 3.17, we seek further algebraic restrictions on $\ker r_{G,n}$.

Referring back to the fundamental exact sequence of pure braid groups defined in Chapter 1, we recall that the homomorphism π_* of Theorem 1.4 was the homomorphism induced by the projection map $\pi : F_{0,n} \to F_{0,n-1}$, where $\pi(z_1,\cdots,z_n^{\bullet}) = (z_1,\cdots,z_{n-1})$. A similar homomorphism exists from $F_{0,n} \to F_{0,n-r}$ which is defined by deleting any subset $(z_{\mu_1},\cdots,z_{\mu_r})$ of the coordinates in the array $(z_1,\cdots,z_n)$. Restricting our attention for the moment to the case $k = 1$, we see that there are n distinct homomorphisms, which we denote by $\pi_*^{(1)},\cdots,\pi_*^{(n)}$ from the pure braid group $P_n = \pi_1 F_{0,n}$ to the pure braid group P_{n-1}, where $\pi_*^{(k)}$ is the homomorphism induced by the projection map $\pi^{(k)}: F_{0,n} \to F_{0,n-1}$ by $\pi^{(k)}(z_1,\cdots,z_{k-1},z_k z_{k+1},\cdots,z_n) = (z_1,\cdots,z_{k-1},z_{k+1},\cdots,z_n)$. In each case $\ker \pi_*^{(k)}$ is a free group of rank $n-1$, which we denote by $F_n^{(k)}$. These n homomorphisms can be interpreted geometrically as "pulling out the kth string" of the braid, or the kth component of the link. In the representation of P_n by automorphisms of the free group they also have the natural meaning: "set x_k equal to 1." In the abstract group P_n generated by the elements $\{A_{ij},\ 1 \leq i < j \leq n\}$ the homomorphism π_*^k has the meaning "set $A_{ij} = 1$ if $i = k$ or $j = k$."

In the Gassner matrix group a similar homomorphism, denoted $((\pi_*^k))$, can be defined by setting the indeterminate t_k in the matrices in $((P_n))^{\mathfrak{a}}$ equal to 1 and deleting the kth row and column. We will show:

THEOREM 3.18. *The diagram:*

$$\begin{array}{ccc} P_n & \xrightarrow{\ \tau_{G,n}\ } & ((P_n))^{\mathfrak{a}} \\ \Big\downarrow \pi_*^k & & \Big\downarrow ((\pi_*^k)) \\ P_{n-1} & \xrightarrow{\ \tau_{G,n-1}\ } & ((P_{n-1}))^{\mathfrak{a}} \end{array}$$

is commutative.

Proof. We prove Theorem 3.18 for $k = n$. The same number of steps suffice for arbitrary k, but the notation is more cumbersome. A weak version of Theorem 3.18 is proved by a different method in [Gassner, 1961].

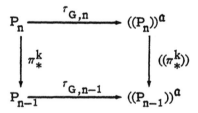

We first show that if $\beta \in \ker \tau_{G,n}$, then $\pi_*^k \beta \in \ker \tau_{G,n-1}$. Let the action of β on the free group F_n be given by $x_i\beta = A_i x_i A_i^{-1}$, $1 \le i \le n$. Since $\beta \in \ker \tau_{G,n}$, it follows from Theorem 3.16 that $A_i \in [\ker \mathfrak{a}_n, \ker \mathfrak{a}_n]$, where $\mathfrak{a}_n$ denotes the abelianizing homomorphism acting on F_n. This implies that:

$$(3\text{-}44) \qquad A_i = \prod_{m=1}^{m_i} [\eta_{im1}, \eta_{im2}] \qquad \eta_{im\ell} \in F_n', \ \ell = 1,2 .$$

Since $\eta_{im\ell} \in F_n'$, it follows that $\eta_{im\ell}$ has exponent sum zero in each letter $x_1,\cdots,x_n$. Therefore if we set $x_n = 1$, each $\eta_{im\ell}$ will go over to a word $\tilde{\eta}_{im\ell}$ in F_{n-1}', and hence A_i will go over to a word $\bar{A}_i \in [\ker \mathfrak{a}_{n-1}, \ker \mathfrak{a}_{n-1}]$, and $\pi_*^n \beta$ maps x_i into $\bar{A}_i x_i \bar{A}_i^{-1}$ for each $1 \le i \le n-1$. It then follows from Theorem 3.16 that $\pi_*^n \beta \in \ker \tau_{G,n-1}$. This tells us that a homomorphism exists from $((P_n))^{\mathfrak{a}}$ to $((P_{n-1}))^{\mathfrak{a}}$, defined unambiguously by $(\tau_{G,n-1})(\pi_*^k)(\tau_{G,n}^{-1})$.

To see that this homomorphism coincides with $((\pi_*^k))$, we consult the basic formula (3-1) for computing free derivatives, and observe that if i and j are both different from n, we will obtain the same result if we either:

(i) Compute $\partial(x_i\beta)/\partial x_j$, and then set $x_n = 1$ in every term of the resulting element of JF_n, or

(ii) Set $x_n = 1$ in the word $x_i\beta \in F_n$, and then compute the partial derivative. Therefore the entries in the $(n-1) \times (n-1)$ matrix $\tau_{G,n-1}\pi_*^n(\beta)$ will be precisely the same as those in the upper $(n-1)\times(n-1)$ box of the matrix obtained from $\tau_{G,n}(\beta)$ by setting $t_n = 1$, hence:

$$(3\text{-}45) \qquad ((\pi_*^k)) = (\tau_{G,n-1})(\pi_*^k)(\tau_{G,n}^{-1}) .$$

To understand this result a little better, we observe that setting $t_n = 1$ in $((P_n))^{\mathfrak{a}}$ gives an obvious homomorphism from $((P_n))^{\mathfrak{a}}$ to a group of matrices over $JF_{n-1}^{\mathfrak{a}}$. Since the entries in the last row of elements in $((P_n))^{\mathfrak{a}}$ are of the form:

$$\delta_{jn}(A_n)^{\alpha} + (1-t_n)\left(\frac{\partial A_n}{\partial x_j}\right)^{\alpha} \tag{3-46}$$

it follows that on setting $t_n = 1$ the last row will go over to a row in which all entries except the nth are zero. Therefore the entries in the upper left $(n-1) \times (n-1)$ box multiply independently of the rest of the matrix, hence if the last row and column are deleted we obtain a group which is a homomorphic image of $((P_n))^{\alpha}$, and in fact has been identified as $((P_{n-1}))^{\alpha}$. ||

An immediate consequence of Theorems 3.17 and 3.18 is:

COROLLARY 3.18.1.

$$\ker \tau_{G,n} \subset (\ker \tau_{G,n-1}) \left(\bigcap_{k=1}^{n} (\ker \pi_*^k \cap R''_{n-1}) \right) .$$

We remark that the group $\bigcap_{k=1}^{n} (\ker \pi_*^k \cap R''_{n-1})$ is non-trivial for every $n \geq 3$. The main value of Corollary 3.18.1, together with Theorem 3.17 is that they give a basis for an inductive argument if one attempts to prove that the Gassner representation is faithful, and otherwise place restrictions on the location of elements in $\ker \tau_{G,n}$.

The method used to prove Theorem 3.18 fails if we attempt to apply it to the Burau representation, because it is not necessarily true that $\pi_*^k \ker \tau_{B,n}$ is included in $\ker \tau_{B,n-1}$. Nevertheless, a weak version of Theorem 3.17 exists for the Burau representation, because the group P_n admits a decomposition (in fact, n different decompositions, corresponding to $k = 1, 2, \cdots, n$) as a semi-direct product of the subgroup P_{n-1} and the normal subgroup $\ker \pi_*^k$. This decomposition goes over to an analogous decomposition in the matrix group $((P_n))^{\psi}$, hence the Burau representation is faithful if and only if:

(i) $\tau_{B,n}$ acts faithfully on the subgroup P_{n-1} of P_n, where P_{n-1} is the subgroup of all braid automorphisms in P_n which map $x_n \to x_n$.

(ii) The matrix groups $\tau_{B,n}(P_{n-1})$ and $\tau_{B,n}(\ker \pi_*^n)$ intersect trivially.

(iii) $\tau_{B,n}$ acts faithfully on $\ker \pi_*^n$.

Property (i) is true if $n = 4$, and could be the basis for an inductive hypothesis for arbitrary n; however, it is not known whether either (ii) or (iii) are true.

One further algebraic restriction on elements in $\ker \tau_{B,n}$ and $\ker \tau_{G,n}$ should be mentioned. If the Burau or Gassner matrix corresponding to an element $\beta \in P_n$ is trivial, it will certainly be trivial for any special value of the indeterminates, e.g., -1. This gives a quick necessary condition. Unfortunately, the subgroup of P_{n-1} which gives the identity matrix under the mapping $t_1 = t_2 = \cdots t_n = t = -1$ is not well understood. A partial characterization is given in Magnus and Peluso, 1969; however, their result is not easily translated into the form needed in the present problem. Our experience is, however, that most elements which satisfy the conditions of Theorems 3.17 and 3.18 also satisfy the requirement that the matrix be trivial if $t = -1$.

Before closing this chapter, we discuss briefly the special case of the group B_4, which is in many ways a considerably easier group to handle than B_n for $n \geq 5$. It seems possible that one or both of our representations are faithful on B_4, but not on B_n if $n \geq 5$. The following holds:

THEOREM 3.19. *The Burau representation of* B_4 *is faithful iff the matrix group generated by the matrices*:

(3-47)
$$\|\sigma_3\sigma_1^{-1}\|^{\psi} = \begin{bmatrix} -t & 1 & 0 \\ 0 & 1 & 0 \\ 0 & 1 & -t^{-1} \end{bmatrix} \quad \|\sigma_2\sigma_3\sigma_1^{-1}\sigma_2^{-1}\|^{\psi} = \begin{bmatrix} 1-t^{-1} & -t^{-1} & t^{-1} \\ 1-t^2 & -t^{-1} & 0 \\ 1 & -t^{-1} & 0 \end{bmatrix}$$

is free and has rank 2.

Proof. By Theorem 3.14: $\ker \tau_{B,4} \subset B'_4$. By results in [Gorin and Lin, 1969] the group B'_4 is a semi-direct of two free groups of rank 2, i.e.,

$$B'_4 = B'_3 \cdot K$$

where:

(3-48) B'_3 = subgroup generated by $\{\sigma_2\sigma_1^{-1}, \sigma_1\sigma_2\sigma_1^{-2}\}$ is free of rank 2

(3-49) K = subgroup generated by $\{\sigma_3\sigma_1^{-1}, \sigma_2\sigma_3\sigma_1^{-1}\sigma_2^{-1}\}$ is free of rank 2 .

Hence the representation will be faithful iff:

(i) The matrix group $\|B'_3\|^{\psi}$ is free

(ii) $\|B_3\|^{\psi} \cap \|K\|^{\psi} = \mathrm{ID}$

(iii) The matrix group $\|K\|^{\psi}$ is free of rank 2.

The matrices defined by equation (3-47) are the generators of $\|K\|^{\psi}$, in the reduced representation given in Theorem 3.11, hence it is only necessary to establish (i) and (ii).

To establish (i), we compute the matrices corresponding to $\sigma_2\sigma_1^{-1}$ and $\sigma_1\sigma_2\sigma_1^{-2}$ in the reduced representation.

(3-50)
$$\|\sigma_2\sigma_1^{-1}\|^{\psi} = \begin{bmatrix} -t^{-1} & t^{-1} & 0 \\ -1 & 1-t & 1 \\ 0 & 0 & 1 \end{bmatrix} \quad \|\sigma_1\sigma_2\sigma_1^{-2}\|^{\psi} = \begin{bmatrix} 0 & -t & 1 \\ t^{-1} & -t^{-1}+1-t & 1 \\ 0 & 0 & 1 \end{bmatrix}$$

Observing that the last row in both of these matrices is (0 0 1), it follows that the upper left 2×2 boxes multiply independently of the remainder of the matrix. Since the matrices in the upper left 2×2 box are precisely those which generate the commutator subgroup B'_3 of B_3 in the Burau representation of B_3, and since these elements generate a free subgroup of B_3 [Gorin and Lin, 1969], it follows from Theorem 3.15 that the representation is faithful on this subgroup. Thus (i) is true.

To establish (ii), suppose $M \in \|\beta_3'\|^{\psi} \cap \|K\|^{\psi}$. Observe that if we replace the indeterminate t in the matrix group $\|B_3'\|^{\psi}$ by -1, then the matrix group goes over to a group of matrices which are still free of rank 2 (cf. proof of Theorem 3.15), hence are still a faithful representation of B_3'. On the other hand, if we replace the indeterminate t by -1 in the group $\|K\|^{\psi}$, then $\|K\|^{\psi}$ goes over to a group of matrices in which the center row is (0 1 0). Hence if $M \in \|B_3'\|^{\psi} \cap \|K\|^{\psi}$, then M must, on setting $t = -1$ and deleting the last row and column, go over to a matrix of the form

$$\begin{bmatrix} a & b \\ 0 & 1 \end{bmatrix}.$$

Since the determinant of this matrix must be 1, it then follows that $a = 1$. Now, our matrix is an element in the matrix group obtained from $\|B_3\|^{\psi}$ by setting $t = -1$, and this group includes the matrix s_1 defined in equation (3.36), and the powers of s_1 are in fact precisely all possible matrices of the form

$$\begin{bmatrix} 1 & b \\ 0 & 1 \end{bmatrix}.$$

Since s_1 is the image of σ_1 in the representation of B_3, it follows that M can only be a power of $\|\sigma_1\|^{\psi}$. But the only power of σ_1 which is in B_3' is the trivial power, hence $M = \text{Id}$. ||

It appears to be a very difficult problem to decide whether a particular group of 3×3 matrices is free, hence Theorem 3.19 does not yield a decision whether the Burau representation of B_4 is faithful.

APPENDIX TO CHAPTER 3: PROOF OF THEOREM 2.10

The reader is referred to Chapter 2 for the statement of Theorem 2.10, and of its relevance to the algebraic link problem. Our proof will be a modification of a proof due to Burau, 1936.

We are given an element $\beta \in B_n$, $n \geq 2$, which is represented by a positive word P. It is assumed that P has letter length m, and also that P involves *every* braid generator $\sigma_1, \cdots, \sigma_{n-1}$. Let $\nabla_\beta(t_1, \cdots, t_\mu)$ be the Alexander polynomial of the link $\hat{\beta}$. We wish to prove that $\nabla_\beta(t_1, \cdots, t_\mu)$ has degree $m-n+1$ and leading coefficient 1. This is equivalent to the assertion that the reduced Alexander polynomial $\tilde{\nabla}_\beta(t) = \nabla_\beta(t, \cdots, t)$ of $\hat{\beta}$ has degree $m-n+1$ and leading coefficient 1.

According to Theorem 3.11, the reduced Alexander polynomial of $\hat{\beta}$ is related to $\det(\|\beta\|_r^\psi - \mathrm{Id})$ by equation (3-28), hence Theorem 2.10 is equivalent to the assertion that $|\,\|\beta\|_r^\psi - \mathrm{Id}|$ has degree m and leading coefficient 1.

Let $\Delta_{i_1 i_2 \cdots i_k}$, $1 \leq k \leq n-2$, be the minor determinant of the matrix which is formed by the terms at the intersection of the i_j^{th} row and column for each $j = 1, \cdots, k$. Then $|\,\|\beta\|_r^\psi - \mathrm{Id}|$ may be expanded as a sum:

$$|\,\|\beta\|_r^\psi - \mathrm{Id}| = \det \|\beta\|_r^\psi + (-1)^{n-1} + \sum$$

where Σ is an appropriate sum of terms of the type $\Delta_{i_1 i_2 \cdots i_k}$, which contains each minor determinant $\Delta_{i_1 i_2 \cdots i_k}$ belonging to each possible unordered array $i_1 i_2 \cdots i_k$ $(1 \leq k \leq n-2)$ exactly once, with coefficient +1 or −1. We observe that it follows from the definitions in (3-26) that $\det \|\beta\|_r^\psi = (-t)^m$, hence our theorem will be true if we can establish that $\Delta_{i_1 i_2 \cdots i_k}$ always has degree strictly *less* than m.

We will establish the latter statement by a double induction on the indices n and m. If $n = 2$, then $\beta = \sigma_1^m$, $m \geq 1$. In this case the reduced Burau matrix $\|\beta\|_r^\psi$ is the 1×1 matrix $\|(-t)^m\|$, and the assertion is trivially true.

Suppose that we are able to establish the assertion for all positive words Q in the group B_n (any fixed n) which have the special property that Q involves *all* of the generators $\sigma_1, \cdots, \sigma_{n-1}$, and *also* that Q involves one of these generators exactly once. Let m_0 be the length of any one such word, say Q_0. (Of course, m_0 might be arbitrarily large.) We assume (inductively) that the assertion is true for all words Q_0' in B_n which has letter length $m > m_0$, and which are formed from Q_0 by the addition of new positive letters. Now consider the word $\sigma_i Q_0'$, $1 \leq i \leq n-1$. Suppose that the reduced Burau matrix for Q_0' is $\|q_{ij}(t)\|$. From equation (3-26) we see that the reduced Burau matrix for $\sigma_i Q_0'$ will differ from that for Q_0' only in the ith row, which will have as its jth entry $(tq_{i-1,j} - tq_{ij} + q_{i+1,j})$ if $i \neq 1$, or $(-tq_{ij} + q_{2j})$ if $i = 1$. It then follows that in going from the minor determinant $\Delta_{i_1 i_2 \cdots i_k}$ of $\|Q_0'\|_r^\psi$ to the corresponding minor determinant of $\|\sigma_i Q_0'\|_r^\psi$ the degree can be increased by at most 1.[11] Since $\det \|Q_0'\|_r^\psi = (-t)^m$ and $\det \|\sigma_i Q_0'\|_r^\psi = (-t)^{m+1}$, we see that

---

[11] The effect of the term $tq_{i-1,j}$ (if $i \neq 1$) on the minor determinant $\Delta_{i_1 i_2 \cdots i_k}$ will be one of the following:

1. If $i-1$ and i are both in the array $i_1 i_2 \cdots i_k$, with, say, $i-1 = i_j$ and $i = i_{j+1}$, then the effect of the term $tq_{i-1,k}$ will be to add the minor determinant $t\Delta_{i_1 \cdots i_{j-1} i_j i_j i_{j+1} \cdots i_k} = 0$.
2. If $i-1$ is not in the array $i_1 i_2 \cdots i_k$, but if i is in this array, with $i = i_j$, then the effect of the term $tq_{i-1,j}$ will be to add the minor determinant $\pm t\Delta_{i_1 \cdots i_{j-1} i_{i-1} i_{i+1} \cdots i_k}$. This minor determinant is necessarily already in the sum Σ, although its degree is now increased by 1, and its sign may be altered.
3. If neither $i-1$ nor i is in the array $i_1 i_2 \cdots i_k$, no new term will be added.

A similar argument applies to the term $q_{i+1,j}$ (if $i \neq 1$) and to the term q_{2j} (if $i = 1$).

the assertion will be true in the general case if we can establish if for all positive words Q which involve all the generators $\sigma_1,\cdots,\sigma_{n-1}$, but involve one of these exactly once.

We now study the class of words Q defined above. Each word Q_0 in the class will be a product of a positive word $W(\sigma_1,\cdots,\sigma_{i-1})$, a positive word $V(\sigma_{i+1},\cdots,\sigma_{n-1})$ and the letter σ_i, in some order. Since cyclic permutation of a positive word does not alter the Alexander polynomial of the associated link, we may without loss of generality assume that the factor σ_i is the first letter in Q_0. Since W and V commute (in B_n), we then have:

$$Q_0 = \sigma_i W(\sigma_1,\cdots,\sigma_{i-1})V(\sigma_{i+1},\cdots,\sigma_{n-1}).$$

Now, $W(\sigma_1,\cdots,\sigma_{i-1}) \in B_i$, while $V(\sigma_{i+1},\cdots,\sigma_{n-1}) \in B_{n-i-1}$ (after an appropriate relabeling of the generators). Suppose that the reduced Burau matrices for W and V (regarded as elements of B_i and B_{n-i-1} respectively) are:

$$\|W\|_r^{\psi} = \begin{bmatrix} w_{11} & \cdots & w_{1,i-1} \\ \cdot & & \cdot \\ \cdot & & \cdot \\ w_{i-1,1} & \cdots & w_{1,i-1} \end{bmatrix}$$

$$\|V\|_r^{\psi} = \begin{bmatrix} v_{i+1,i+1} & \cdots & v_{i+1,n-1} \\ \cdot & \cdot & \cdot \\ \cdot & & \cdot \\ v_{n-1,i+1} & \cdots & v_{n-1,n-1} \end{bmatrix}$$

Going back to the definition of the reduced Burau matrix, as given in Lemma 3.11.1, we then find that the reduced Burau matrix for $Q_0 = \sigma_i WV$ will be

$$\|Q_0\|_r^{\psi} = \begin{bmatrix} w_{11} & \cdots & w_{1,i-1} & \cdots & w_{1,i} & \cdots & 0 & \cdots & 0 \\ \cdot & & \cdot & & \cdot & & \cdot & & \\ \cdot & & \cdot & & \cdot & & \cdot & & \\ w_{i-1,1} & & w_{i-1,i-1} & & w_{i-1,i} & & 0 & \cdots & 0 \\ tw_{i-1,1} & & tw_{i-1,i-1} & & tw_{i-1,i}-t+v_{i+1,i} & & v_{i+1,i+1} & \cdots & v_{i+1,n-1} \\ 0 & & 0 & & v_{i+1,i} & & v_{i+1,i+1} & \cdots & v_{i+1,n-1} \\ \cdot & & \cdot & & \cdot & & \cdot & & \cdot \\ \cdot & & \cdot & & \cdot & & \cdot & & \cdot \\ 0 & & 0 & & v_{n-1,i} & & v_{n-1,i+1} & & v_{n-1,n-1} \end{bmatrix}$$

We are now ready to prove our assertion about the degree of the minor determinants of the reduced Burau matrices for positive words Q having the special property described earlier. The assertion is known to be true for $n = 2$ and for all $m \geq 1$. We assume inductively that it is true for words Q in the groups B_d, whenever $d < n$. Now consider the word Q_0 in B_n, where the reduced Burau matrix of Q_0 is as given above. Let $\Delta_{i_1 i_2 \cdots i_k}$ be any minor determinant of Q_0. If i is *not* in the array $i_1 i_2 \cdots i_k$, then our minor determinant $\Delta_{i_1 i_2 \cdots i_k}$ is a product of minor determinants in B_i and B_{n-i-1}, hence by the induction hypothesis its degree is less than the degree of

$$\det \|Q_0\|_r^{\psi} = (-t)\cdot \det \|W\|_r^{\psi} \cdot \det \|V\|_r^{\psi} .$$

If $i-1$, i and $i+1$ are all in the array, then $(t \cdot (i-1)^{st}$ row $+ (i+1)^{st}$ row) may be added to the ith row to reduce the minor determinant to one which is exactly t times a product of minor determinants in B_i and B_{n-i-1}, and again the inductive hypothesis completes the proof. Finally, if $i-1$ and i are in the array, but $i+1$ is not (or if i and $i+1$ are in the array, but $i-1$ is not) then we may subtract t times the $(i-1)^{st}$ row from the ith (or subtract the $(i+1)^{st}$ row from the ith) to reduce the situation once more to the case where the inductive hypothesis may be applied. This completes the proof of Theorem 2.10. ||

CHAPTER 4

MAPPING CLASS GROUPS

Let T_g denote a closed orientable surface of genus g; and let $z_1^0, \cdots, z_n^0$ denote n fixed but arbitrarily chosen points on T_g. Recall that in Chapter 1 the symbol $\pi_1 F_{0,n} T_g$ denoted the pure braid group on T_g (with base point $(z_1^0, \cdots, z_n^0)$) and $\pi_1 B_{0,n} T_g$ denoted the full braid group on T_g (with the same base point). The following similar notation is meant to suggest relationships to be developed in this chapter:

$\mathcal{F}_n T_g$ denotes the group of all orientation preserving homeomorphisms $h : T_g \to T_g$ such that, for each i, $h(z_i^0) = z_i^0$.

$\mathcal{B}_n T_g$ denotes the group of all orientation preserving homeomorphisms $h : T_g \to T_g$ such that $h(\{z_1^0, \cdots, z_n^0\}) = \{z_1^0, \cdots, z_n^0\}$.

These two groups are to be endowed with the compact-open topology.

$\pi_0(\mathcal{F}_n T_g, \mathrm{id})$ denotes the group (!) of path components of $\mathcal{F}_n T_g$ and is called the *pure mapping class group* of T_g.

$\pi_0(\mathcal{B}_n T_g, \mathrm{id})$ denotes the group of path components of $\mathcal{B}_n T_g$ and is called the *(full) mapping class group* of T_g. The notation $M(g, n)$ is also used for this group.

Note that $\mathcal{F}_0 T_g = \mathcal{B}_0 T_g$ and $\mathcal{F}_1 T_g = \mathcal{B}_1 T_g$. Hence $M(g, 0) = \pi_0(\mathcal{F}_0 T_g) = \pi_0(\mathcal{B}_0 T_g)$ and $M(g, 1) = \pi_0(\mathcal{F}_1 T_g) = \pi_0(\mathcal{B}_1 T_g)$.

This chapter will be devoted to such problems as determining the structure of and finding generators and relations for the groups $M(g, n) = \pi_0(\mathcal{B}_n T_g, \mathrm{id})$. In brief outline, the topics covered are the following:

1) For each pair of integers $g, n \geq 0$ there is a natural homomorphism

$$j_* : M(g,n) \to M(g,0)$$

induced by the inclusion map $j: \mathcal{B}_n T_g \subset \mathcal{B}_0 T_g$. If $g \geq 2$, then $\ker j_*$ is isomorphic to the n-string braid group $\pi_1 B_{0,n} T_g$ of the surface T_g. If $g = 1$, $n \geq 2$ or $g = 0$, $n \geq 3$ then $\ker j_*$ is isomorphic to $\pi_1 B_{0,n} T_g$/center. In the special case $g = 0$, this result allows one to compute presentations for the mapping class groups $M(0,n)$.

2) The group $M(g,0)$ is generated by a finite set of "twist" maps (known variously as Dehn twists or Lickorish twists!).

3) The space $T_{g,0}$ may be regarded as a ramified covering of the punctured sphere. If conditions are right, then homeomorphisms of the punctured sphere lift to homeomorphisms of $T_{g,0}$. Even more, entire classes of maps on the punctured sphere lift (modulo covering transformations) to entire classes of maps on $T_{g,0}$. This allows one to find large subgroups of $M(g,0)$.

We begin in Section 4.1 by studying the homomorphism i_* from $\pi_0 \mathcal{F}_n T_g$ to $\pi_0 \mathcal{F}_0 T_g$ which is induced by the inclusion map. It is established in Theorem 4.2 that the kernel of i_* is the n-string pure braid group $\pi_1 F_{0,n} T_g$ of T_g modulo its center. The identification of $\ker i_*$ is made explicit, by means of the *evaluation map* defined in Section 4.1. The establishment of this relationship between braid groups and mapping class groups is a fundamental result, which will be shown to have far-reaching consequences in determining certain algebraic and structural properties of the groups $\pi_0 \mathcal{F}_n T_g$, $M(0,n)$, and even $M(g,0)$. The analogous result for the full mapping class group $M(g,n) = \pi_0 \mathcal{B}_n T_g$ is given in Theorem 4.3.

Section 4.2 is concerned with the group $M(0,n)$, the mapping class group of the n-punctured sphere. Using the results of Section 4.1, we obtain generators and defining relations for $M(0,n)$ (see Theorem 4.5). This group is very closely related to Artin's braid group, and again to the braid group $\pi_1 B_{0,n} S^2$ of the n-punctured sphere. We note that the system of relations obtained in Theorem 4.5 was discovered earlier, by a very different method, by W. Magnus [see Magnus, 1934]. The proof given here appeals to us because it shows the relationship between Magnus' earlier results (which did not appear to generalize to surfaces other than S^2) and the larger theory of braid groups of surfaces, as presented in Chapter 1.

Section 4.3 contains a summary of the essential features of W. B. R. Lickorish's proof that $M(g, 0)$ is generated by a particular finite set of "twist" maps (Theorem 4.6). We remark on the fact that Lickorish's proof is essentially geometric in nature, as was the earlier proof of the same result by [Dehn, 1938]. Although it has been known for some time that the group $M(g, 0)$ is canonically isomorphic to $\text{Aut}\,\pi_1 T_g / \text{Inn}\,\pi_1 T_g$ [see Mangler, 1939], it seems to be an extremely difficult problem to establish Theorem 4.6 algebraically. The heart of the difficulty appears to be twofold: (i) it is difficult to characterize the class of elements in $\pi_1 T_g$ which have simple representatives [see D. Chillingworth, 1969] and (ii) algebraicists have not succeeded in finding an appropriate replacement for counting arguments based on intersections of curves on surfaces.

Section 4.4 contains an example of the application of the technique of lifting and projecting homeomorphisms, which may be used to escalate known results about the mapping class group of the n-punctured sphere, by applying them to obtain information about the mapping class groups of appropriate branched covering surfaces of S^2. In particular, we use this technique in Theorem 4.7 and 4.8 to obtain defining relations for the group $M(2, 0)$, by establishing that $M(2, 0)$ is a finite extension of $M(0, 6)$. The proof of Theorem 4.7 is essentially the proof given in [Birman and Hilden, 1973], adapted to the special case $g = 2$, $n = 0$.

Finally, Section 4.5 contains a brief survey of the present state of knowledge with respect to mapping class groups of surfaces.

4.1. *The natural homomorphism from* $M(g, n)$ *to* $M(g, 0)$

DEFINITION. Recall that $z_1^0, \cdots, z_n^0$ are a set of n distinguished points on T_g, and that if $\mathcal{k} \in \mathcal{F}_n T_g$, then $\mathcal{k}(z_i^0) = z_i^0$ for each $i = 1, \cdots, n$. Recall also that, in Chapter 1, the array $(z_1^0, \cdots, z_n^0)$ was chosen to be the base point for the space $F_{0,n}T_g$. Using these points, we now define (for each pair of integers $n, g \geq 0$) an *evaluation map*:

$$\epsilon_{gn} : \mathcal{F}_0 T_g \to F_{0,n}T_g = \{(p_1, \cdots, p_n) | p_i \in T_g;\ p_i \neq p_j \text{ if } i \neq j\} \tag{4-1}$$

by $\epsilon_{gn}(h) = (h(z_1^0), \cdots, h(z_n^0))$. Observe that ϵ_{gn} is continuous with the given topologies on $\mathcal{F}_0 T_g$ (compact open topology) and $F_{0,n}T_g$ (subspace topology for $F_{0,n}T_g \subset T_g \times \cdots \times T_g$).

THEOREM 4.1 [Birman, 1969b]. *The evaluation map* $\epsilon_{gn} : \mathcal{F}_0 T_g \to F_{0,n}T_g$ *is a locally trivial fibering with fibre* $\mathcal{F}_n T_g$.[1]

Proof. Note that $\mathcal{F}_n T_g$ is a closed subgroup of the topological group $\mathcal{F}_0 T_g$ and that two elements h and g of $\mathcal{F}_0 T_g$ have the same image under ϵ_{gn} if and only if they are in the same left coset of $\mathcal{F}_n T_g$ in $\mathcal{F}_0 T_g$. This observation results in a natural identification of $F_{0,n}T_g$ with the quotient space $\mathcal{F}_0 T_g / \mathcal{F}_n T_g$, and this identification is easily seen to be a homeomorphism which turns ϵ_{gn} into a projection map and exhibits $F_{0,n}T_g$ as a homogeneous space.

By [Hu, *Homotopy Theory*, Exercise D4, p. 99], Theorem 4.1 will follow immediately once it has been shown that, relative to ϵ_{gn}, there is a local cross section of the homogeneous space $F_{0,n}T_g$ in $\mathcal{F}_0 T_g$ at the single point $\vec{z}^0 = \epsilon_{gn}(\mathcal{F}_n T_g) = (z_1^0, \cdots, z_n^0) \in F_{0,n}T_g$ (i.e., there is a neighborhood $U(\vec{z}^0)$ of $\vec{z}^0$ in $F_{0,n}T_g$ and a map $\mathcal{X} : U(\vec{z}^0) \to \mathcal{F}_0 T_g$ such that

[1] We observe that the statement and proof of Theorem 4.1 are very similar to the statement and proof of Theorem 1.2, Chapter 1. Just as in Chapter 1, the pure mapping class group (respectively pure braid group) is a little easier to work with than the full mapping class group (full braid group). The full mapping class group (respectively full braid group) is an extension of the pure mapping class group (pure braid group) by the symmetric group Σ_n.

$\epsilon_{gn}\mathcal{X} = \text{id}$). Choose pairwise disjoint Euclidean neighborhoods $U(z_1^0),\cdots,U(z_n^0)$ of $z_1^0,\cdots,z_n^0$, respectively, on T_g. Then $U(\vec{z}^0) = \{(u_1,\cdots,u_n)|u_i \in U(z_i^0)\}$ is a neighborhood of $\vec{z}^0$ in $F_{0,n}T_g$. Construct a family of homeomorphisms $\{f_u \in \mathcal{F}_0T_g | \vec{u} \in U(\vec{z}^0)\}$, depending continuously on $\vec{u}$, such that, for each $\vec{u} \in U(\vec{z}^0)$, $f_u(z_i^0) = u_i$ and $f_u|(T_g - \bigcup_{i=1}^{n} U(z_i^0)) =$ identity. Define $\mathcal{X}(\vec{u}) = f_u$. ||

COROLLARY 4.1.1. *There is an exact sequence of homotopy groups*

$$\text{(4-2)} \longrightarrow \pi_1\mathcal{F}_0T_g \xrightarrow{\epsilon_{gn_*}} \pi_1F_{0,n}T_g \xrightarrow{d_{gn_*}} \pi_0\mathcal{F}_nT_g \xrightarrow{i_{gn_*}} \pi_0\mathcal{F}_0T_g \longrightarrow \pi_0F_{0,n}T_g = 1\ .$$

Proof. The homomorphism ϵ_{gn_*} of Corollary 4.1.1 is induced by the evaluation map ϵ_{gn} the homomorphism i_{gn_*} by the inclusion $\mathcal{F}_nT_g \to \mathcal{F}_0T_g$. The exact sequence is simply the exact homotopy sequence of the fibering $\epsilon_{gn}: \mathcal{F}_0T_g \to F_{0,n}T_g$. ||

(Remark: The surjection $i_* = i_{gn_*}$ is the analogue for the pure mapping class groups of the homomorphism $j_* = j_{gn_*}: M(g,n) \to M(g,0)$ of full mapping class groups named in the title of this paragraph and studied below in Theorem 4.3.)

For simplicity, we will replace the symbols i_{gn_*}, ϵ_{gn_*}, d_{gn_*} by i_*, ϵ_*, d_* respectively. Corollary 4.1.1 becomes effective only when i_* has been structurally determined by a careful examination of $\text{im}\, d_* = \ker i_*$ and $\ker d_*$:

THEOREM 4.2 [Birman, 1969b]. *For each pair of integers* $g,n \geq 0$ *let* $i_{gn_*} = i_*: \pi_0\mathcal{F}_nT_g \to \pi_0\mathcal{F}_0T_g$ *be the homomorphism induced by inclusion. Then* $\ker i_* =$ *image* $d_* \approx \pi_1F_{0,n}T_g$ *if* $g \geq 2$. *If* $g = 1$, $n \geq 2$ *or* $g = 0$, $n \geq 3$ *then* $\ker i_* \approx \pi_1F_{0,n}T_g/\text{center}$.

Proof. We consider the final segment of the long exact sequence of Corollary 4.1.1.

Except for the case $g = 1$, whose proof will be omitted, the theorem is an immediate consequence of Lemma 4.2.1 below (which shows that $\ker d_* \subset \text{center } \pi_1 F_{0,n}T_g$) and of Lemmas 4.2.2-4.2.4 (which identify center $\pi_1 F_{0,n}T_g$ explicitly). Before these lemmas are stated and proved, however, the construction of d_* and the proof of the relationship $\ker i_* = \operatorname{im} d_*$ will be recalled.

Construction of d_*. Suppose $\beta \in \pi_1 F_{0,n}T_g$, with β represented by a loop $(\beta_1, \cdots, \beta_n): I \to F_{0,n}T_g$. Then it is an easy matter to construct an isotopy $h_t: T_g \to T_g$ $(0 \le t \le 1)$ such that $h_0 = \text{id}$, $h_t(x_i) = \beta_i(t)$, and hence $h_1 \in \mathcal{F}_n T_g$. Indeed, the construction is obvious for generators $\{A_{ij} | 1 \le i < j \le n\}$ of $\pi_1 F_{0,n}T_g$ (cf. Chapter 1) and by composition of isotopies can be extended to all elements of $\pi_1 F_{0,n}T_g$. Then $[h_1] = d_*\beta$.

ker $i_* =$ *im* d_*. This follows immediately from the fact that the sequence (4-2) is exact. It will be of interest to obtain image d_* explicitly. Suppose that $h \in \mathcal{F}_n T_g$ with $[h] \in \ker i_*$. Since $h \in \mathcal{F}_n T_g$, h fixes $z_1^0, \cdots, z_n^0$ pointwise. Since $[h] \in \ker i_*$, h is isotopic to the identity map on T_g, say by an isotopy h_t $(0 \le t \le 1)$, $h_0 = \text{identity}$, $h_1 = h$. Then $(h_t(z_1^0), \cdots, h_t(z_n^0))$ represents an element β of $\pi_1 F_{0,n}T_g$ and $d_*\beta = [h]$. ||

LEMMA 4.2.1. $\ker d_* \subset \text{center } \pi_1 F_{0,n}T_g$.

Proof. Suppose $\alpha \in \ker d_* = \operatorname{im} \epsilon_*$, and let $H \in \pi_1 \mathcal{F}_0 T_g$ be such that $\epsilon_* H = \alpha$. The element H is represented by a loop $h = \{h_t | 0 \le t \le q\}$ in $\mathcal{F}_0 T_g$, where each h_t is in $\mathcal{F}_0 T_g$ and $h_0 = h_1 = \text{id}$. Then $\epsilon(h_t) = (h_t(x_1), \cdots, h_t(x_n))$ $(0 \le t \le 1)$ represents α. Let $\beta \in \pi_1 F_{0,n}T_g$, with β represented by $(\beta_1(s), \cdots, \beta_n(s))$ $(0 \le s \le 1)$. Define $G: I \times I \to F_{0,n}T_g$ by $G(t,s) = (h_t\beta_1(s), \cdots, h_t\beta_n(s))$ $((t,s) \in I \times I)$. Then G is continuous and $G|\partial(I \times I)$ represents the homotopy class $\alpha\beta\alpha^{-1}\beta^{-1}$. Since β was an arbitrary element of $\pi_1 F_{0,n}T_g$, we may conclude that $\alpha \in \text{center } \pi_1 F_{0,n}T_g$. ||

LEMMA 4.2.2. *If* $g \ge 2$, *then* $\text{center } \pi_1 F_{0,n}T_g = 1$.

Proof. Recall (Theorem 1.4) the exact sequence (1-7)

$$1 \longrightarrow \pi_1 F_{n-1,1} T_g \xrightarrow{\partial_*} \pi_1 F_{0,n} T_g \xrightarrow{\pi_*} \pi_1 F_{0,n-1} T_g \longrightarrow 1$$

of pure braid groups. If $n = 1$, $\pi_1 F_{0,1} T_g = \pi_1 T_g$ which is centerless. Assume inductively that $\pi_1 F_{0,n-1} T_g$ is centerless. Since π_* is surjective, $\pi_*(\text{center } \pi_1 F_{0,n} T_g) \subset \text{center } \pi_1 F_{0,n-1} T_g = 1$. Hence center $\pi_1 F_{0,n} T_g$ lies in the group $\text{im } j_* = \ker \pi_*$. But $\pi_1 F_{n-1,1} T_g \approx \text{im } j_*$ is a free group of rank > 1, hence centerless. Thus center $\pi_1 F_{0,n} T_g = 1$.

The next two lemmas treat the case $g = 0$, i.e., $T_g = S^2$. ||

LEMMA 4.2.3 [Gillette and Van Buskirk, 1968; the proof given here is, however, different]. *Let* $\delta_1, \cdots, \delta_{n-1}$ *be standard generators of* $\pi_1 B_{0,n} S^2$, $n \geq 3$ (cf. *Theorem 1.10*). *Then the center of* $\pi_1 B_{0,n} S^2$ *is the subgroup of order 2 generated by* $(\delta_1 \delta_2 \cdots \delta_{n-1})^n$.

Proof. Observe that the mapping ν from $\pi_1 B_{0,n} S^2$ to Σ_n defined by $\nu(\delta_i) = (i, i+1)$, $1 \leq i \leq n-1$ is a homomorphism. Since Σ_n is centerless for $n \geq 3$, it follows that any element in the center of $\pi_1 B_{0,n} S^2$ must have the identity permutation. Hence it suffices to identify the center of the subgroup $D_n S^2$ of $\pi_1 B_{0,n} S^2$ where $D_n S^2$ is defined to be the subgroup of all elements in $\pi_1 B_{0,n} S^2$ whose associated permutation leaves the letter n invariant.

Coset representatives for $D_n S^2$ in $\pi_1 B_{0,n} S^2$ may be chosen to be the n elements 1 and $\delta_{n-1} \delta_{n-2} \cdots \delta_k$, $1 \leq k < n-1$. Application of the Schreier-Reidemeister method (cf. Lemma 1.8.2) yields the following group presentation for $D_n S^2$ (details of the calculation are left to the reader):

generators: $\delta_1, \delta_2, \cdots \delta_{n-2}$

(4-3) *relations*: $\delta_i \delta_j = \delta_j \delta_i \quad |i-j| \geq 2$

(4-4) $\delta_i \delta_{i+1} \delta_i = \delta_{i+1} \delta_i \delta_{i+1}$

(4-5) $(\delta_1 \delta_2 \cdots \delta_{n-2})^{2(n-1)} = 1$.

Comparing this group presentation with the group presentation obtained in Theorem 1.4.1 for $\pi_1 B_{0,n} E^2$, we find that $D_n S^2$ is a quotient group of $\pi_1 B_{0,n-1} E^2$, the kernel being the normal closure of the element $(\sigma_1 \sigma_2 \cdots \sigma_{n-2})^{2(n-1)}$ in $\pi_1 B_{0,n-1} E^2$. But, by Corollary 1.8.4 of Chapter 1, the element $(\sigma_1 \sigma_2 \cdots \sigma_{n-2})^{n-1}$ generates the infinite cyclic center of $\pi_1 B_{0,n-1} E^2$. Since adding to a group presentation a relation which kills a power of an element in the center cannot introduce any new commutativity relations into the group, it follows that the center of $D_n S^2$ must be generated by $(\delta_1 \delta_2 \cdots \delta_{n-2})^{n-1}$ and have order 1 or 2. But, now, as a consequence of relations (1-27), (1-28), (1-29) one finds:[2]

$$(4\text{-}6) \qquad (\delta_1 \delta_2 \cdots \delta_{n-1})^n = (\delta_1 \delta_2 \cdots \delta_{n-2})^{n-1} (\delta_{n-1} \cdots \delta_2 \delta_1^2 \delta_2 \cdots \delta_{n-1})$$
$$= (\delta_1 \delta_2 \cdots \delta_{n-2})^{n-1} .$$

This implies that the center of $D_n S^2$ (= center $\pi_1 B_{0,n} S^2$) is the subgroup of order 1 or 2 generated by $(\delta_1 \delta_2 \cdots \delta_{n-1})^n$.

To see that the element $(\delta_1 \delta_2 \cdots \delta_{n-1})^n$ cannot have order 1 in $\pi_1 B_{0,n} S^2$, consider first the case where n is odd. Since relators (1-27), (1-28), (1-29) have exponent sum $0, 0, 2(n-1)$ respectively, but $(\delta_1 \delta_2 \cdots \delta_{n-1})^n$ has exponent sum $n(n-1)$, it follows that if n is odd this element cannot be a relator. The proof for n even follows easily by utilizing the exact sequence (1-7) of Theorem 1.4, observing that $(\delta_1 \delta_2 \cdots \delta_{n-1})^n \in \pi_1 F_{0,n} S^2$ and that its image in $\pi_1 F_{0,n-1} S^2$ under the homomorphism π_* is $(\delta_1 \delta_2 \cdots \delta_{n-2})^{n-1}$. ||

LEMMA 4.2.4. *If* $n \geq 3$, *then*

$$\text{center } \pi_1 B_{0,n} S^2 \subset \text{center } \pi_1 F_{0,n} S^2 \subset \ker d_* .$$

---

2 In the derivation of (4-6) we used the fact that $\pi_1 B_{0,n} S^2$ admits an automorphism: $\delta_i \to \delta_{n-i}$, $1 \leq i \leq n-1$.

Proof. By Lemma 4.2.3,

$$\text{center } \pi_1 B_{0,n} S^2 \subset \text{center } \pi_1 F_{0,n} S^2$$

is the cyclic subgroup of order 2 generated by $(\delta_1 \cdots \delta_{n-1})^n$, hence Lemma 4.2.4 will be true if we can prove that $d_*(\delta_1 \cdots \delta_{n-1})^n = 1$. In the standard geometric model of $\pi_1 F_{0,n} E^2$, $(\sigma_1 \cdots \sigma_{n-1})^n$ can be pictured as in Figure 13a and 13b (drawn for the case $n = 4$). Then as a motion of points on S^2, $(\delta_1 \cdots \delta_{n-1})^n$ can be pictured as in Figure 13c.

Recall now the construction of d_*. Without loss of generality, $z_1^0, \cdots, z_n^0$ are spaced at equal distances on a longitude joining z_1^0 at the equator with z_n^0 at the north pole (as in Figure 13c), and the motion of points pictured in Figure 13c can be realized by a rotation $h_t : S^2 \to S^2$ $(0 \le t \le 1)$ about the axis joining the north and south pole, $h_0 = h_1 = \mathrm{id}$. Then $d_*(\delta_1 \cdots \delta_{n-1})^n) = [h_1] = 1$. ||

Lemmas 4.2.1 - 4.2.4 complete the proof of Theorem 4.2 except for the case $g = 1$. The proof for $g = 1$ will not be included here. One proceeds as in the other cases to identify center $\pi_1 B_{0,n} T_1$ (see Birman, 1969a). The center is a free abelian group of rank 2 (isomorphic to $\pi_1 T_1$). ||

THEOREM 4.3. *For each pair of integers* $g, n \ge 0$ *let* $j_* = j_{gn_*} : M(g,n) \to M(g,0)$ *be the homomorphism induced by the inclusion* $j : \mathcal{B}_n T_g \subset \mathcal{B}_0 T_g$. *Then* $\ker j_*$ *is isomorphic to* $\pi_1 B_{0,n} T_g$ *for* $g \ge 2$. *If* $g = 1$, $n \ge 2$ *or* $g = 0$, $n \ge 3$, *then* $\ker j_*$ *is isomorphic to* $\pi_1 B_{0,n} T_g/\text{center}$.

Proof. The proof is essentially a repetition of the arguments used to prove Theorems 4.1 and 4.2. As in Theorem 4.1, we may establish that there is an evaluation map $\epsilon'_{gn} : \mathcal{B}_0 T_g \to B_{0,n} T_g$, which is a locally trivial fibering with fibre $\mathcal{B}_n T_g$. As in Corollary 4.1.1, this fibering defines an exact sequence

$$(4\text{-}7) \quad \longrightarrow \pi_1 \mathcal{B}_0 T_g \xrightarrow{\epsilon'_{gn_*}} \pi_1 B_{0,n} T_g \xrightarrow{d'_{gn_*}} M(g,n) \xrightarrow{j_{gn_*}} M(g,0) \longrightarrow \pi_0 B_{0,n} T_g = 1 .$$

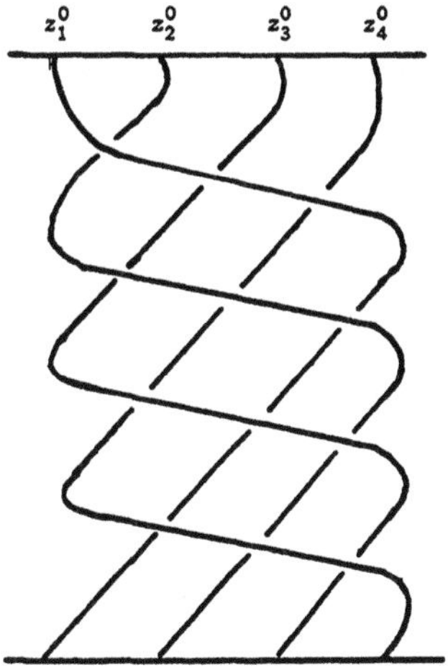

The geometric braid
$(\sigma_1 \sigma_2 \sigma_3)^4 \in \pi_1 F_{0,4} E^2$

(a)

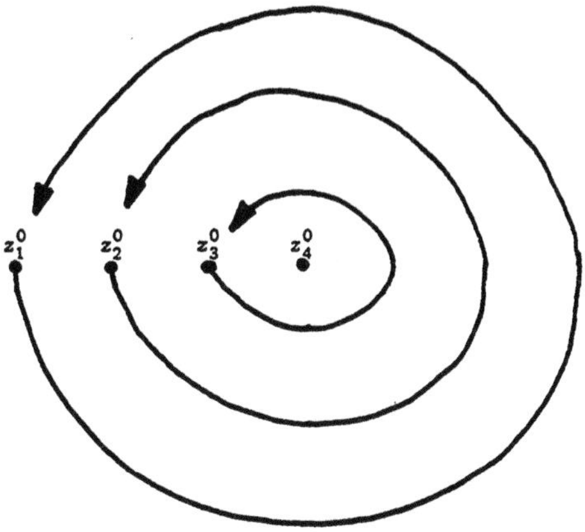

$(\sigma_1 \sigma_2 \sigma_3)^4$ as a motion of points on E^2

(b)

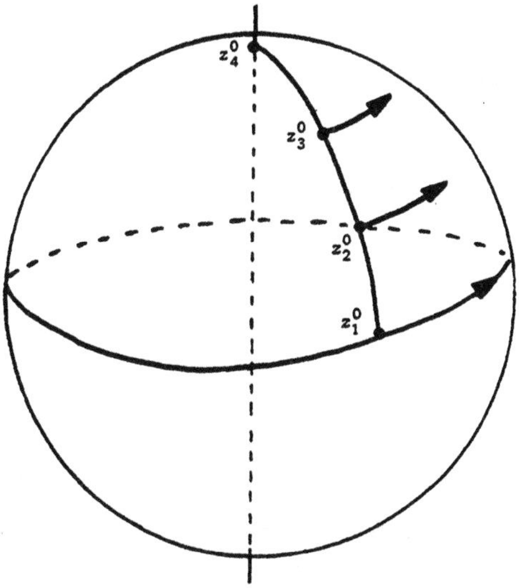

$(\delta_1 \delta_2 \delta_3)^4$ as a motion of points on S^2

(c)

Fig. 13.

As in the proof of Theorem 4.2, we may establish that $\ker j_{gn_*}$ is naturally isomorphic to $\pi_1 B_{0,n} T_g$ if $g \geq 2$, $n \geq 2$, or to $\pi_1 B_{0,n} T_g/\text{center}$ if $g = 1$, $n \geq 2$ or $g = 0$, $n \geq 3$. ||

We close this section by making explicit the situations described by Theorems 4.2 and 4.3. First consider the case $n = 1$, $g \geq 2$. In this case the homomorphism i_{g1_*} and j_{g1_*} of Theorems 4.2 and 4.3 coincide (because $\mathcal{F}_n T_g = \mathcal{B}_n T_g$ if $n = 0$ or 1), and moreover $\pi_1 F_{0,1} T_g = \pi_1 B_{0,1} T_g \approx \pi_1 T_g$, so that our theorems imply that $\ker i_{g1_*} = \ker j_{g1_*}$ is isomorphic to $\pi_1 T_g$. We will now describe in an explicit fashion how to find generators for the subgroup $\ker j_{g1_*}$ of the mapping class group $M(g, 1) = \pi_0 \mathcal{F}_1 T_g = \pi_0 \mathcal{B}_1 T_g$.

Let z_1^0 be the base point for $\pi_1 T_g$, and suppose that c is any simple closed curve on $\pi_1 T_g$ which contains the base point z_1^0. Let N be a cylindrical neighborhood of c on T_g, parametrized by (y, θ), with $-1 \leq y \leq +1$, $0 \leq \theta \leq 2\pi$, where the curve c is described by $y = 0$, and the base point z_1^0 by $(0, 0)$. We now define a map $h_{cz_1} : T_g \to T_g$ by the rule that if a point is in N, then its image is given by:

$$h_{cz_1} : (y, \theta) \to (y, \theta + 2\pi y) \text{ if } \quad 0 \leq y \leq 1 \qquad (4\text{-}8)$$
$$(y, \theta) \to (y, \theta - 2\pi y) \text{ if } -1 \leq y \leq 0 ,$$

while all points of $T_g - N$ are left fixed. We call such a map a *spin of* z_1^0 *about* c. (See Figure 14.)[3] Note that $h_{cz_1} \in M(g, 1)$.

Now let $a_1, \cdots, a_g$, $b_1, \cdots, b_g$ be 2g simple closed curves on T_g, meeting in the base point z_1^0 but otherwise disjoint, and having the property that their homotopy classes generate $\pi_1 T_g$, and also that the

3 The reader who is familiar with Dehn twists, to be discussed in Section 4.3, will recognize that a spin of z about c may also be described as a product of a pair of Dehn twists, in opposite directions, about curves c_1 and c_2 defined by, say, $y = +\frac{1}{2}$ and $y = -\frac{1}{2}$ respectively.

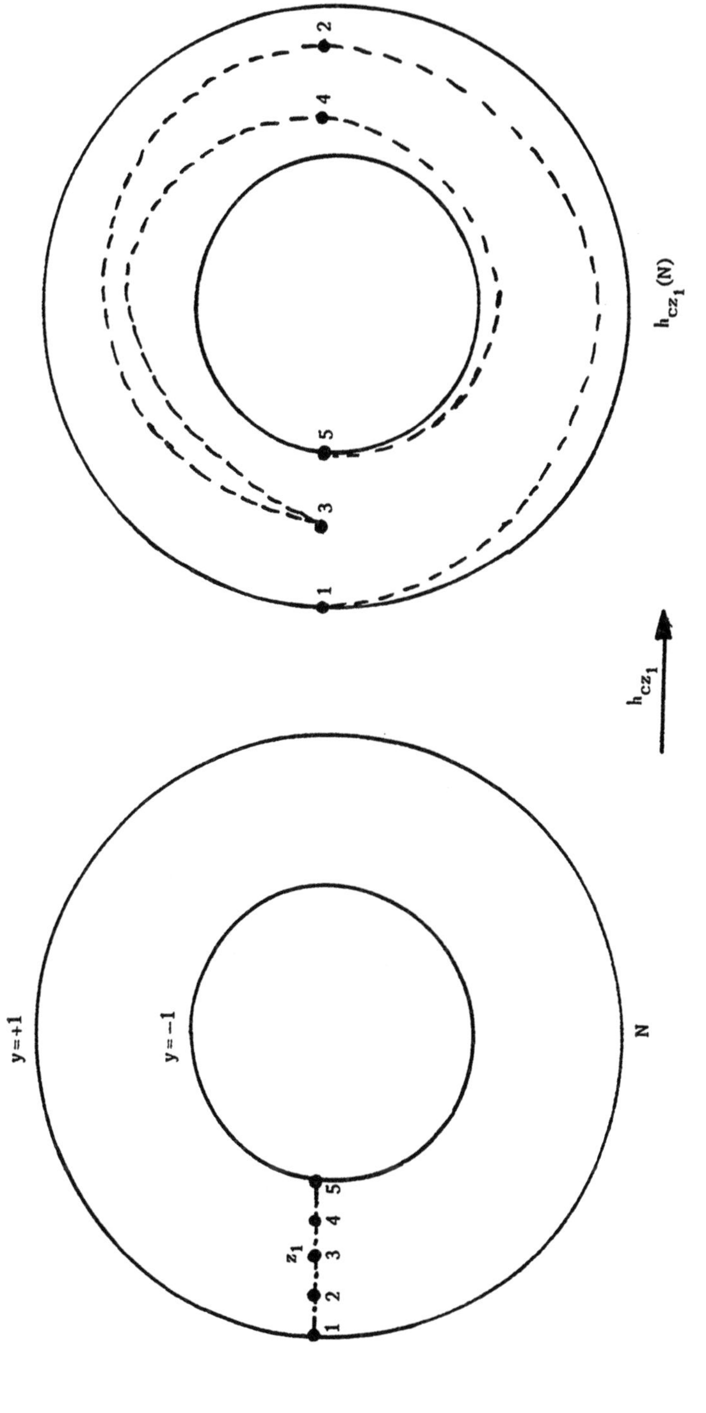

Fig. 14.

homotopy class of $a_1b_1a_1^{-1}b_1^{-1}\cdots a_gb_ga_g^{-1}b_g^{-1}$ is trivial. Then the isotopy classes of the spin maps $\mathscr{h}_{a_1z_1},\cdots,\mathscr{h}_{a_gz_1},\mathscr{h}_{b_1z_1},\cdots,\mathscr{h}_{b_gz_1}$ on $T_{g,1}$ generate $\ker j_{g,1_*}$, and the isotopy class of the product

$$\mathscr{h}_{a_1z_1}\mathscr{h}_{b_1z_1}\mathscr{h}_{a_1z_1}^{-1}\mathscr{h}_{b_1z_1}^{-1}\cdots\mathscr{h}_{a_gz_1}\mathscr{h}_{b_gz_1}\mathscr{h}_{a_gz_1}^{-1}\mathscr{h}_{b_gz_1}^{-1}$$

is trivial.

In the case where n is arbitrary, the single point z_1^0 which serves as base point for $\pi_1T_g = \pi_1F_{0,1}T_g = \pi_1B_{0,n}T_g$ (regarded as a subgroup of $\pi_0\mathcal{F}_1T_g = \pi_0\mathcal{B}_1T_g$) is replaced by an array $(z_1^0,\cdots,z_n^0)$ which determines a base point for $\pi_1F_{0,n}T_g$, and also for $\pi_1B_{0,n}T_g$. Let $\{a_{ij}, b_{ij};\ 1\le i\le g,\ 1\le j\le n\}$ be $2gn$ simple closed curves on T_g, where $a_{i1}=a_i$ and $b_{i1}=b_i$ are as previously defined, and where each curve a_{ij} (respectively b_{ij}) is freely homotopic to a_{i1} (respectively b_{i1}) on T_g, and each curve a_{ij} (respectively b_{ij}) contains the base point z_j^0, but no other point z_k^0 if $k\neq j$. Then the isotopy classes of the spin maps

$$\{\mathscr{h}_{a_{ij}z_j},\mathscr{h}_{b_{ij}z_j};\ 1\le i\le g,\ 1\le j\le n\}$$

generate $\ker i_{gn_*}$. To obtain a set of generators for $\ker j_{gn_*}$ one adds to this set any set of maps on the surface T_g which generate the full group of permutations of the points $(z_1^0,\cdots,z_n^0)$; for example, enclose each pair (z_j^0,z_{j+1}^0) in a disc D_j which avoids all points z_k^0 $(k\neq j)$ and map T_g to itself by a map $\mathscr{h}_j$ which fixes $T_g - D_j$ pointwise, and interchanges z_j^0 and z_{j+1}^0 "nicely."

The case $g=1$ is similar to that described above. The case $g=0$ will be treated separately in Section 4.2, below.

Remark. The exact sequences (4-2) and (4-7) offer information also about the higher homotopy groups of $\mathcal{F}_nT_g$, $\mathcal{F}_0T_g$; $\mathcal{B}_nT_g$, and $\mathcal{B}_0T_g$. Information about these higher homotopy groups appears in Quintas 1968; McCarty 1963; and various papers by M. E. Hamstrom, who has calculated these groups for a number of difficult cases.

4.2. *The mapping class group of the n-punctured sphere*

THEOREM 4.4. *Every orientation-preserving self-homeomorphism of a 2-sphere, or of a 2-sphere with one point removed, is isotopic to the identity map. Thus* $M(0,0) = M(0,1) = 1$.

Remark. While Theorem 4.4 is a well-known folk theorem, the author was surprised to find that no published proof seemed to exist. We therefore fill the gap by including a pleasant proof, due to J. H. Roberts, who has kindly located it after a gap of 40 years, communicated it, and allowed us to use it!

Proof. We first establish that $M(0,0)$ is isomorphic to $M(0,1)$. This may be seen by examining the exact sequences (4-2) and (4-7), which coincide when $n = 1$:

$$\to \pi_1 B_{0,1}S^2 \to M(0,1) \to M(0,0) \to \pi_0 B_{0,1}S^2 \ .$$

The space $B_{0,1}S^2 = S^2 - \text{pt.}$ is arcwise-connected, hence $\pi_0 B_{0,1}S^2 = 1$. Also, $\pi_1 B_{0,1}S^2 = \pi_1 S^2 = 1$. Hence $M(0,1) \approx M(0,0)$.

The proof that $M(0,1) = 1$, i.e., that every orientation-preserving homeomorphism of S^2 which fixes a point $p \in S^2$ is isotopic to the identity map via an isotopy which keeps p fixed at each stage, will depend on Lemmas 4.4.1 - 4.4.2 below.

LEMMA 4.4.1 [Alexander, 1923b]. *If* $g: D^n \to D^n$ *is a homeomorphism from the unit n-ball to itself which fixes the (n−1)-sphere* $S^{n-1} = \partial D^n$ *pointwise, then* g *is isotopic to the identity under an isotopy which fixes* S^{n-1} *pointwise. If* $g(0) = 0$, *then the isotopy may be chosen to fix* 0.

Proof. Recall that $D^n = \{x \in E^n / \text{dist}(x,0) = |x| \le 1\}$ and that $S^{n-1} = \{x \in E^n / |x| = 1\}$. Each point $x \in E^n - \{0\}$ has a unique "polar" representation $x = (r, \theta)$, where $r = |x|$ and $\theta = \frac{x}{|x|} \in S^{n-1}$. Suppose that $g(r,\theta) = [R(r,\theta), \Theta(r,\theta)], (r \le 1)$. Then $R(1,\theta) = 1$ and $\Theta(1,\theta) = \theta$ since $g|S^{n-1} =$ identity. Extend $g = (R, \Theta)$ to all of E^n by the identity. Define the isotopy $g_t : E^n \to E^n$ $(0 < t \leq 1; r \leq 1)$ by the formula

$$g_t(r, \theta) = [t \cdot R(r/t, \theta), \Theta(r/t, \theta)] .$$

Observe that $g_1 = g$, $g_t(r, \theta) = (r, \theta)$ for $r > t$, and, for fixed $r \leq t$, the map g_t is the map g with a scale factor. Thus the effect of the isotopy is to push the distorted region of E^n into an n-ball of smaller and smaller radius. The maps g_t approach the identity map continuously as $t \to 0$. Hence g_0 may be defined to be the identity and g_t $(0 \leq t \leq 1)$ is the desired isotopy. Note that if $g(0) = 0$, then $g_t(0) = 0$ for each t. ||

LEMMA 4.4.2 (Schoenflies Theorem). *If* $\mathcal{J}_1$ *and* $\mathcal{J}_2$ *are simple closed curves in the plane* E^2 *and* $h : \mathcal{J}_1 \to \mathcal{J}_2$ *is any homeomorphism, then there is an extension* h_* *of* h *which takes* $\mathcal{J}_1 \cup \mathrm{Int}\, \mathcal{J}_1$ *homeomorphically onto* $\mathcal{J}_2 \cup \mathrm{Int}\, \mathcal{J}_2$.

This lemma will be assumed without proof. For this and other results in plane topology, the reader is referred to the following very nice elementary accounts:

M. H. A. Newman, *Elements of the Topology of Plane Sets of Points*, 2nd ed., Cambridge Univ. Press, Cambridge, 1951.

G. T. Whyburn, *Topological Analysis*, Princeton Math. Series, no. 23, Princeton Univ. Press, Princeton, N. J., 1958. ||

Now, suppose that $p \in S^2$ and that $f : S^2 \to S^2$ is an orientation preserving homeomorphism which fixes p. We must show that there is an isotopy between f and the identity which fixes p at each stage.

If $\mathcal{J}$ is a simple closed curve in $S^2 - \{p\}$, let $\mathrm{Int}\, \mathcal{J}$ denote the component of $S^2 - \mathcal{J}$ which contains p, $\mathrm{Ext}\, \mathcal{J}$ the other component. Choose simple closed curves $\mathcal{J}_1$ and $\mathcal{J}_2$ in $S^2 - \{p\}$ such that $\mathcal{J}_1 \cup f(\mathcal{J}_1) \subset \mathrm{Int}\, \mathcal{J}_2$.

Let $\mathcal{A}_1, \mathcal{A}_2$, and $\mathcal{A}_3$ be disjoint arcs with $(\mathcal{A}_1 \cup \mathcal{A}_2 \cup \mathcal{A}_3) \subset [\mathrm{Int}\, \mathcal{J}_2 - (\mathcal{J}_1 \cup \mathrm{Int}\, \mathcal{J}_1)]$, each of the arcs joining the simple closed curves $\mathcal{J}_1$ and $\mathcal{J}_2$. Let $\{x_i\} = \mathcal{A}_i \cap \mathcal{J}_1$ and $\{y_i\} = \mathcal{A}_i \cap \mathcal{J}_2$ $(i = 1, 2, 3)$. Since f is orientation preserving and fixes p, there are disjoint arcs $\mathcal{K}_1, \mathcal{K}_2$, and $\mathcal{K}_3$ joining $\mathfrak{f}(x_1)$ and y_1, $\mathfrak{f}(x_2)$ and y_2, and $\mathfrak{f}(x_3)$ and y_3, respectively such that $\mathcal{K}_i \subset \mathrm{Int}\, \mathcal{J}_2 - [f(\mathcal{J}_1) \cup \mathrm{Int}\, f(\mathcal{J}_1)]$. (See Figure 15.)

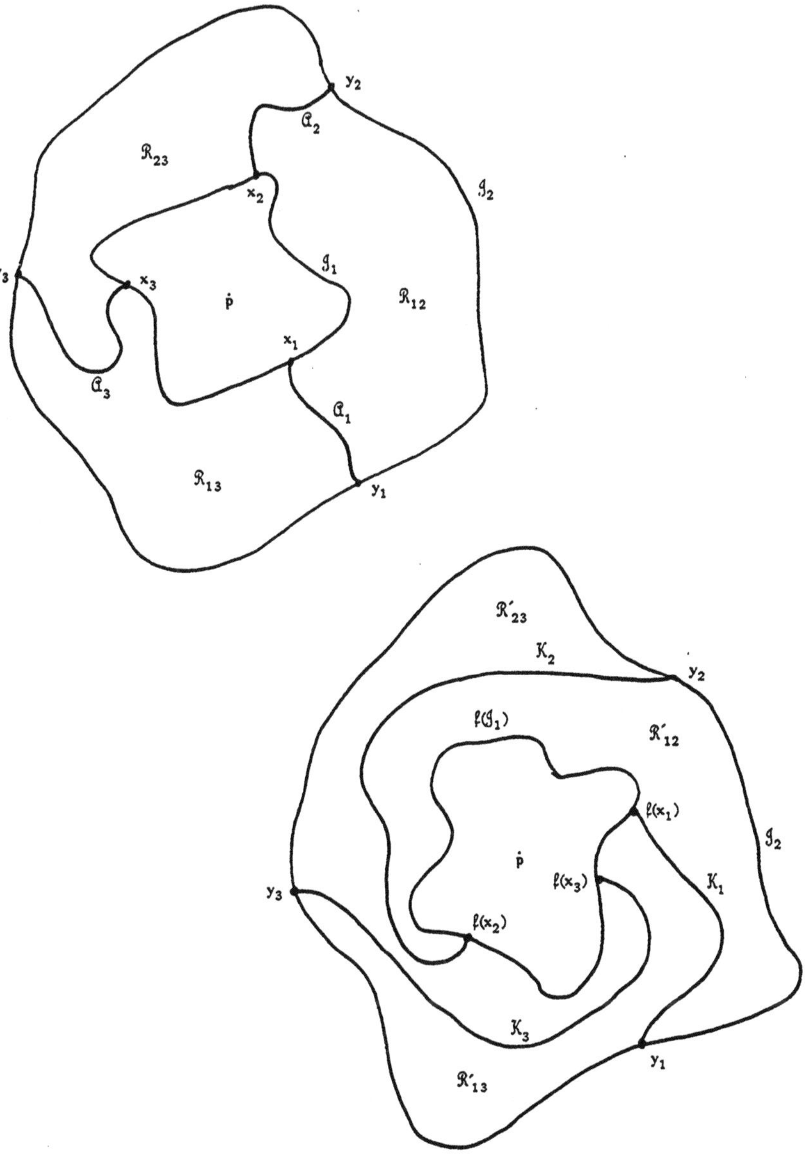

Fig. 15.

Let $\mathcal{R}_{ij}(i \neq j)$ be the component of $\operatorname{Int} \mathcal{J}_2 - [\mathcal{A}_1 \cup \mathcal{A}_2 \cup \mathcal{A}_3 \cup \mathcal{J}_1 \cup \operatorname{Int} \mathcal{J}_1]$ which contains $\mathcal{A}_i \cup \mathcal{A}_j$ in its boundary. Let $\mathcal{R}'_{ij}$ be the component of $\operatorname{Int} \mathcal{J}_2 - [\mathcal{K}_1 \cup \mathcal{K}_2 \cup \mathcal{K}_3 \cup f(\mathcal{J}_1) \cup \operatorname{Int} f(\mathcal{J}_1)]$ which contains $\mathcal{K}_i \cup \mathcal{K}_j$ in its boundary.

It is an easy exercise to construct a homeomorphism $g: S^2 \to S^2$ which has the following properties:

(i) $g|\mathcal{J}_1 \cup \operatorname{Int} \mathcal{J}_1 = f|\mathcal{J}_1 \cup \operatorname{Int} \mathcal{J}_1$

(ii) $g|\mathcal{J}_2 \cup \operatorname{Ext} \mathcal{J}_2 = \text{identity}|\mathcal{J}_2 \cup \operatorname{Ext} \mathcal{J}_2$

(iii) $g(\mathcal{R}_{ij}) = \mathcal{R}'_{ij}$.

Indeed, (i) and (ii) define $g|\operatorname{Bd}\mathcal{A}_i$ $(i = 1,2,3)$. This partial map can be extended to take $\mathcal{A}_i$ to $\mathcal{K}_i$ homeomorphically. This defines $g|\operatorname{Bd}\mathcal{R}_{ij}$ for each $i \neq j$. And by Lemma 4.4.2, $g|\operatorname{Bd}\mathcal{R}_{ij}: \operatorname{Bd}\mathcal{R}_{ij} \to \operatorname{Bd}\mathcal{R}'_{ij}$ can be extended to a homeomorphism $g|\overline{\mathcal{R}}_{ij}: \overline{\mathcal{R}}_{ij} \to \overline{\mathcal{R}}'_{ij}$ (where $\overline{\mathcal{R}}$ denotes the closure of the region $\mathcal{R}$).

By (i) and Lemma 4.4.1, the homeomorphisms f and g are isotopic under an isotopy which moves points only in $\operatorname{Ext} \mathcal{J}_1$. By (ii) and Lemma 4.4.1, the homeomorphisms g and identity are isotopic under an isotopy which moves points only in $\operatorname{Int} \mathcal{J}_2$. Since $g(p) = p$, it follows from the last sentence of Lemma 4.4.1 that the latter isotopy may be chosen to fix p. The composition of the two isotopies gives the desired isotopy between f and the identity map. This completes the proof of Theorem 4.4. ||

THEOREM 4.5. *If* $n \geq 2$, *then* $M(0,n) = \pi_0 \mathcal{B}_n S^2$ *admits a presentation with generators* $\omega_1, \cdots, \omega_{n-1}$ *and defining relations*:

(4-9) $$\omega_i \omega_j = \omega_j \omega_i \qquad |i-j| \geq 2$$

(4-10) $$\omega_i \omega_{i+1} \omega_i = \omega_{i+1} \omega_i \omega_{i+1}$$

(4-11) $$\omega_1 \cdots \omega_{n-2} \omega_{n-1}^2 \omega_{n-2} \cdots \omega_1 = 1$$

(4-12) $$(\omega_1 \omega_2 \cdots \omega_{n-1})^n = 1\,.$$

If $n = 0$ *or* 1, *then* $M(0,n) = 1$.

Proof. The proof for $n = 0$ or 1 is given in Theorem 4.4. For $n = 2$ we note that the map ω_1 which interchanges the punctures on the surface is necessarily non-trivial. To see that its square must be trivial, we show that the group $\pi_0\mathcal{F}_2S^2 = 1$. This follows from the long exact sequence (4-1), using the facts that $\pi_1 F_{0,2}S^2 = 1$ (Theorem 1.10) and $\pi_0 F_{0,2}S^2 = 1$ (because $F_{0,2}$ is arcwise-connected).

If $n \geq 3$, then by Theorem 4.3 the group $M(0, n)$ is an extension of $\pi_1 B_{0,n}S^2/\text{center}$ by $M(0, 0)$. By Theorem 4.4 the group $M(0, 0)$ is trivial, hence $M(0, n)$ is isomorphic to $\pi_1 B_{0,n}S^2/\text{center}$. A presentation for $\pi_1 B_{0,n}S^2$ is given in Theorem 1.10, in terms of generators $\delta_1, \cdots, \delta_{n-1}$. The center of $\pi_1 B_{0,n}S^2$ is identified in terms of these generators in Lemma 4.2.3. Let ω_i denote the image of δ_i under the homomorphism from $(\pi_1 B_{0,n}S^2/\text{center})$ to $M(0, n)$. Then $M(0, n)$ admits the indicated presentation. ||

The generators $\omega_1, \cdots, \omega_{n-1}$ of $M(0, n)$ have a very simple geometric interpretation. Let $q_1, \cdots, q_n$ be the points which are to be deleted from the surface S^2. Enclose each pair (q_i, q_{i+1}) in a disc D_i, chosen so that D_i avoids q_j $(j \neq i, i+1)$. Map S^2 to itself by a map h_i which fixes $(S^2 - D_i)$ pointwise and interchanges q_i and q_{i+1} "nicely." Then $h_i \in M(0, n)$; also h_i is isotopic to the identity on the closed sphere. Moreover, if h_t is the isotopy taking h_i to the identity map, then the orbit $(h_t(q_1), \cdots, h_t(q_n))$ represents the element δ_i in $\pi_1 B_{0,n}S^2$. Thus we can identify the isotopy class of h_i with ω_i.[4]

4.3. *Generators for* M(g, 0)

In this section our focus will be on the group $M(g, 0)$, and on the problem of determining generators for $M(g, 0)$.

Let c be a simple closed curve on $T_{g,0}$, and consider a neighborhood N of c which is homeomorphic to a cylinder. For convenience we will assume that the cylinder is parametrized by coordinate (y, θ), where

---

[4] The reader who is familiar with Dehn twists, to be discussed in Section 4.3, will recognize that h_i may be interpreted as a Dehn twist about a curve c which contains the points q_i and q_{i+1} at $(0,0)$ and $(0,\pi)$ respectively (see e.g. 4-13).

the y axis is the axis of the cylinder, and $-1 \leq y \leq +1$. Also, c is defined by $y = 0$. Define a map $\mathscr{g}_c$ of the cylinder $\to$ itself by:

$$\mathscr{g}_c(y, \theta) = (y, \theta + \pi(y+1)) . \tag{4-13}$$

Then $\mathscr{g}_c$, extended by the identity map to all of $T_{g,0}$, is known as a *Dehn twist* about c. (See Figure 16.) We will establish that the isotopy classes of Dehn twists generate $M(g, 0)$.

A property of Dehn twists is:

LEMMA 4.6.1. *If* p *and* m *are simple closed curves on* T_g, *and* $h_i : T_g \to T_g$ *is an isotopy, with* h_0 = *identity and* $h_1(p) = m$, *then there is an isotopy between* $\mathscr{g}_p$ *and* $\mathscr{g}_m$.

Proof of Lemma 4.6.1. From the definition of a twist map we have $\mathscr{g}_m = h_1 \mathscr{g}_p h_1^{-1}$, hence $h_t \mathscr{g}_p h_t^{-1}$ is an isotopy between $\mathscr{g}_p$ and $\mathscr{g}_m$. ||

We remark that the sense of a Dehn twist depends on the choice of a coordinate system (i.e., an orientation) on N, but that it does not depend on the orientation of the curve c itself. For convenience in drawing pictures we would of course like to choose orientations for the neighborhoods of arbitrary curves on T_g which agree with a given orientation on T_g. To accomplish this, we will adopt the following rule. Assume that T_g is oriented. Note first that if p is a path which crosses the simple closed curve c at a finite set of points, say $a_1, \cdots, a_r$, then the effect on p of a Dehn twist about c will be to break p at each point a_i and insert a copy of c at the break. This will be assumed to be done in such a way that if some very small segment $\Delta p \subset p$ which is close to c is given a local orientation *toward* c, then points on this segment will be moved to the *right* when a positive Dehn twist about c is applied, and to the left if a negative twist is applied. Note that this rule will be valid for directed segments Δp on either side of c, as long as the orientation is always toward c. (Cf. Figure 16; c is the curve $y = 0$.)

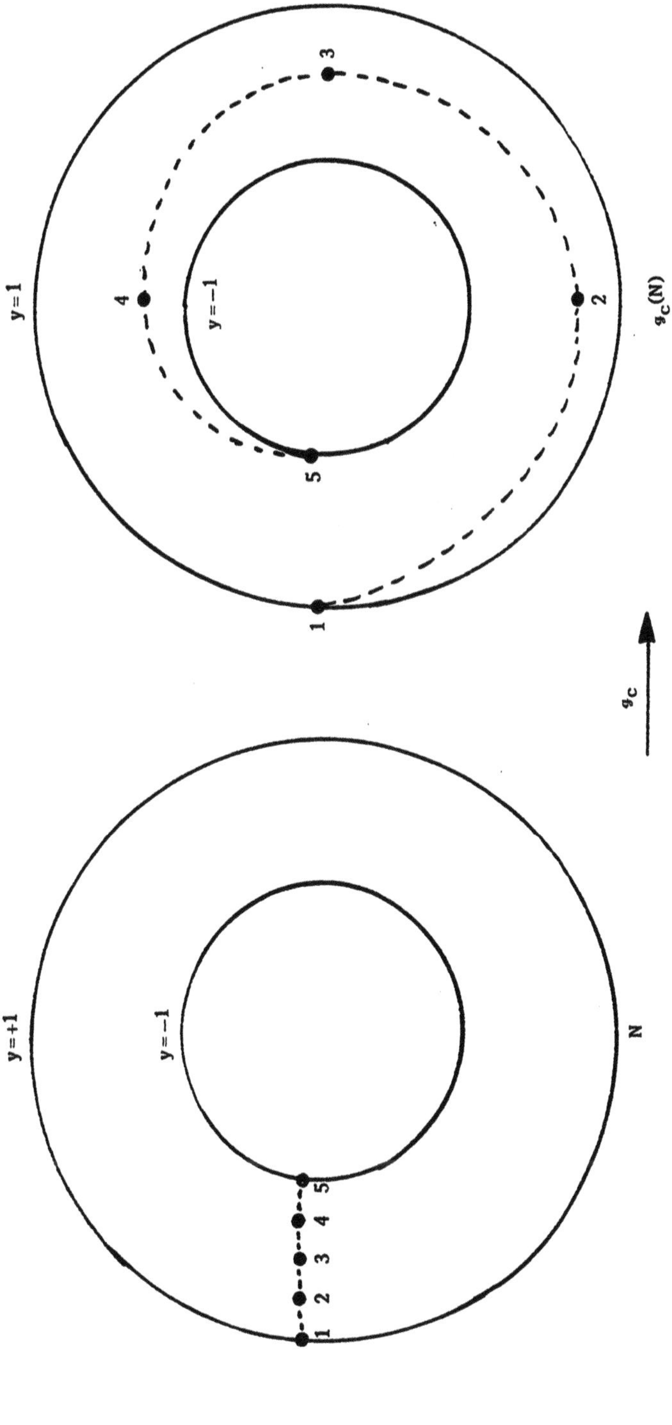

Fig. 16.

THEOREM 4.6 [Dehn, 1938; Lickorish 1962, 1964, 1966].

(i) *Every orientation-preserving homeomorphism of* $T_{g,0} \to T_{g,0}$ *is isotopic to a product of Dehn twists.*

(ii) *Let* $c_1, \cdots, c_{3g-1}$ *be the simple closed curves on* $T_{g,0}$ *which are illustrated in Figure 17. Then a Dehn twist about an arbitrary simple closed curve on* $T_{g,0}$ *is isotopic to a power product of twists about* $c_1, \cdots, c_{3g-1}$.

We will give a detailed proof of (i), and a summary of the main ideas in the proof of (ii). Our method will be based on Lickorish's methods, however we will be able to shorten his proofs by utilizing the results of Theorem 4.3. All homeomorphisms and isotopies are piecewise linear; all paths are polygonal.

Notation:

$T_g = T_{g,0}$

$\mathcal{G}$ group of all orientation-preserving piecewise-linear homeomorphisms of $T_g \to T_g$

$\mathcal{Y}$ subgroup of $\mathcal{G}$ which is generated by Dehn twists

$h, g, \cdots$ elements of $\mathcal{G}$

$g \cong h$ if g is isotopic to h

$\mathcal{D} = \{d \in \mathcal{G} / d \cong \mathrm{id}\}$

$c, p, m, \cdots$ simple closed curves on T_g

$g_c, g_p, g_m, \cdots$ Dehn twists about $c, p, n, \cdots$

$p \sim m$ if $\exists\, h \in \mathcal{G} \ni h(m) = p$

$p \approx m$ if $\exists\, d \in \mathcal{D} \ni d(m) = p$

$p \sim_c m$ if $\exists\, y \in \mathcal{Y} \ni y(m) \approx p$

$|p \cap m|$ cardinality of $p \cap m$

$a, b, d, \cdots$ points on T_g.

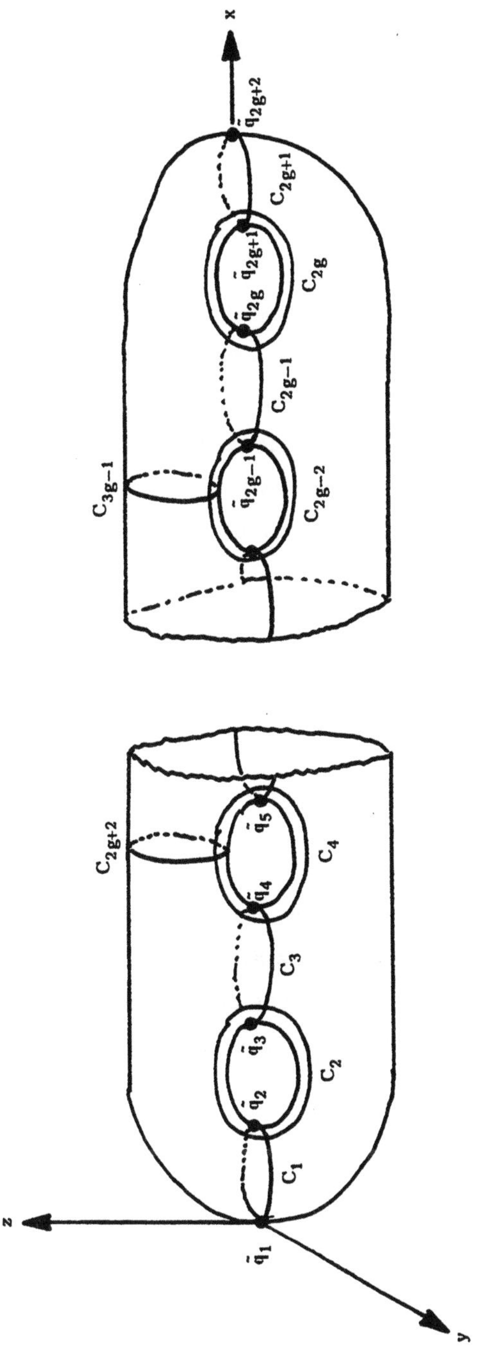
x
$\tilde{q}_{2g+2}$
C_{2g+1}
$\tilde{q}_{2g+1}$
$\tilde{q}_{2g}$
C_{2g}
C_{2g-1}
C_{3g-1}
$\tilde{q}_{2g-1}$
C_{2g-2}
C_{2g+2}
$\tilde{q}_5$
C_4
$\tilde{q}_4$
C_3
$\tilde{q}_3$
C_2
$\tilde{q}_2$
C_1
$\tilde{q}_1$
z
y

Fig. 17.

The general plan of the proof of part (i) will be to examine the image, p, of a particular "meridian" curve m_1 of T_g (see Figure 21) under an arbitrary orientation-preserving homeomorphism h of T_g. It will be shown that one can always find an appropriate power product, say h_1, of Dehn twists, and an isotopy, say d_1 of T_g, such that $d_1 h_1(p)$ does not intersect any of the meridian curves $m_1, \cdots, m_g$. This allows us to find a further sequence of Dehn twists, say h_2, and a second isotopy, d_2, such that $d_2 h_2 d_1 h_1(p) = m_1$. Since $p = h(m_1)$, it will then follow that $h_* = d_2 h_2 d_1 h_1 h$ may be regarded as a homeomorphism of a surface of genus $g-1$, with two discs removed which is obtained by cutting along m_1. Induction on genus, together with the results of Theorem 4.3, will complete the proof. The proof of part (ii) will involve a second, more refined induction on intersection numbers of curves.

Since we wish to apply twist maps to reduce the number of intersections between particular simple, closed curves on T_g, we begin by developing several lemmas (Lemmas 4.6.2 - 4.6.5) which will indicate how to do this:

LEMMA 4.6.2. *Let* p_1 *and* p_2 *be simple closed paths on* T_g, *with* $|p_1 \cap p_2| = 1$. *Then* $p_1 \sim_c p_2$.

Proof of Lemma 4.6.2. Let a be the point $p_1 \cap p_2$. (Figure 18a.) Apply the Dehn twist $\mathfrak{t}_{p_2}$ to p_1. This has the effect of breaking p_1 at a and inserting a copy of p_2 at the break (see Figure 18b). Now apply a Dehn twist $\mathfrak{t}_{p_1}$. Its effect[5] on $\mathfrak{t}_{p_2}(p_1)$ is indicated in Figure 18c, which shows that $\mathfrak{t}_{p_1}\mathfrak{t}_{p_2}(p_1) \approx p_2$, hence $p_1 \sim_c p_2$. ||

[5] In view of Lemma 4.6.1, we may twist about a curve p_1' which is close to p_1, instead of about p_1. We choose p_1' in such a way that $|p_1' \cap \mathfrak{t}_{p_2}(p_1)| = 1$. This makes the picutre a little easier to draw.

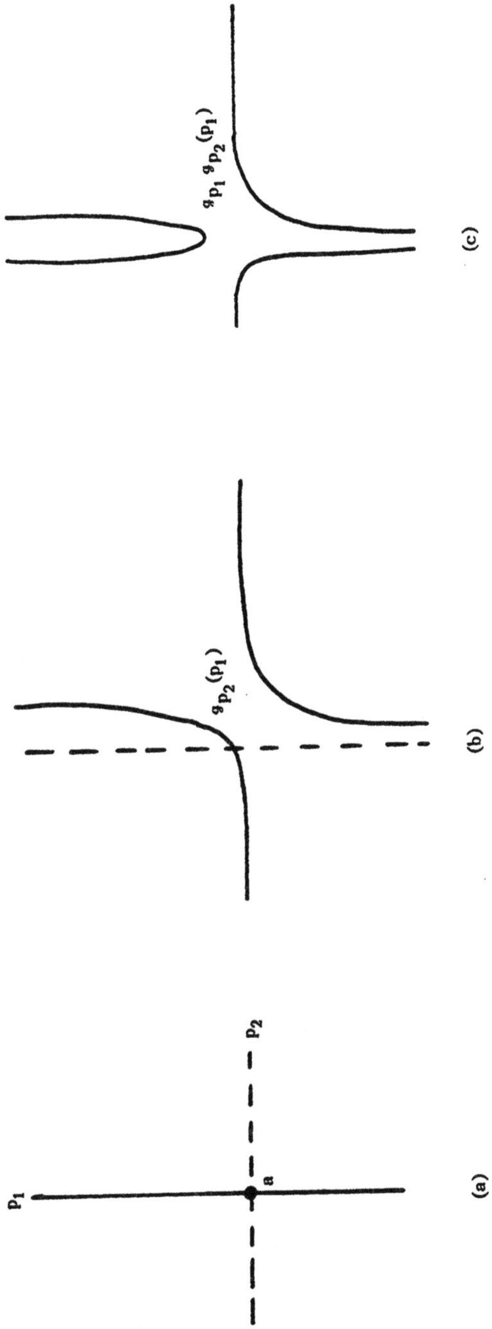

Fig. 18

DEFINITION. Two paths p and m will be said to *meet twice with zero algebraic intersection* if $|p \cap m| = 2$ and if it is possible to orient p in such a way that p has different directions with respect to a given orientation of m at the two points of $p \cap m$. We will indicate this situation by writing $|p \cap m| = 2_0$.

Our next lemma is the central step in the proof of Theorem 4.6, part (i).

LEMMA 4.6.3. *Let* p *and* m *be simple paths on* T_g, *with* p *closed. Let* N *be a neighborhood of* m. *Then there exists a path* u *on* T_g *such that*

1. $u \sim_c p$
2. $u \subset p \cup N$
3. $|u \cap m| = 0$ *or* 2_0.

Proof of Lemma 4.6.3. The proof is by induction on $r = |p \cap m|$. If $r = 0$ or 2_0, take $u = p$. If $r = 1$, and m is closed, we may apply Lemma 4.6.2. If m is not closed, it is clear that we may move p off m by an isotopy which is the identity outside $p \cup N$. Assume, inductively, that the lemma is true if $r < k$. Orient p and m. There are two cases to consider.

Case 1. The set $p \cap m$ contains two adjacent points (on m), say a and b, which are such that p is oriented in the same direction at a as it is at b with respect to the orientation of m (see Figure 19(a)). Choose points a' and b' in N, close to a and b, and as illustrated in Figure 19(a). Let c be a simple path which starts at a' and proceeds, close to p, to the point b', then crosses m once in N to return to a'. Figure 19(b) shows $\mathscr{g}_c(p)$. Then we may find an element $d \in \mathfrak{D}$ such that $d\mathscr{g}_c(p) \subset p \cup N$, and $|d\mathscr{g}_c(p) \cap m| < k$, as in Figure 19(c).[6] The induction hypothesis may now be applied to construct u.

---

6 Note that $|c \cap m| < |p \cap m|$.

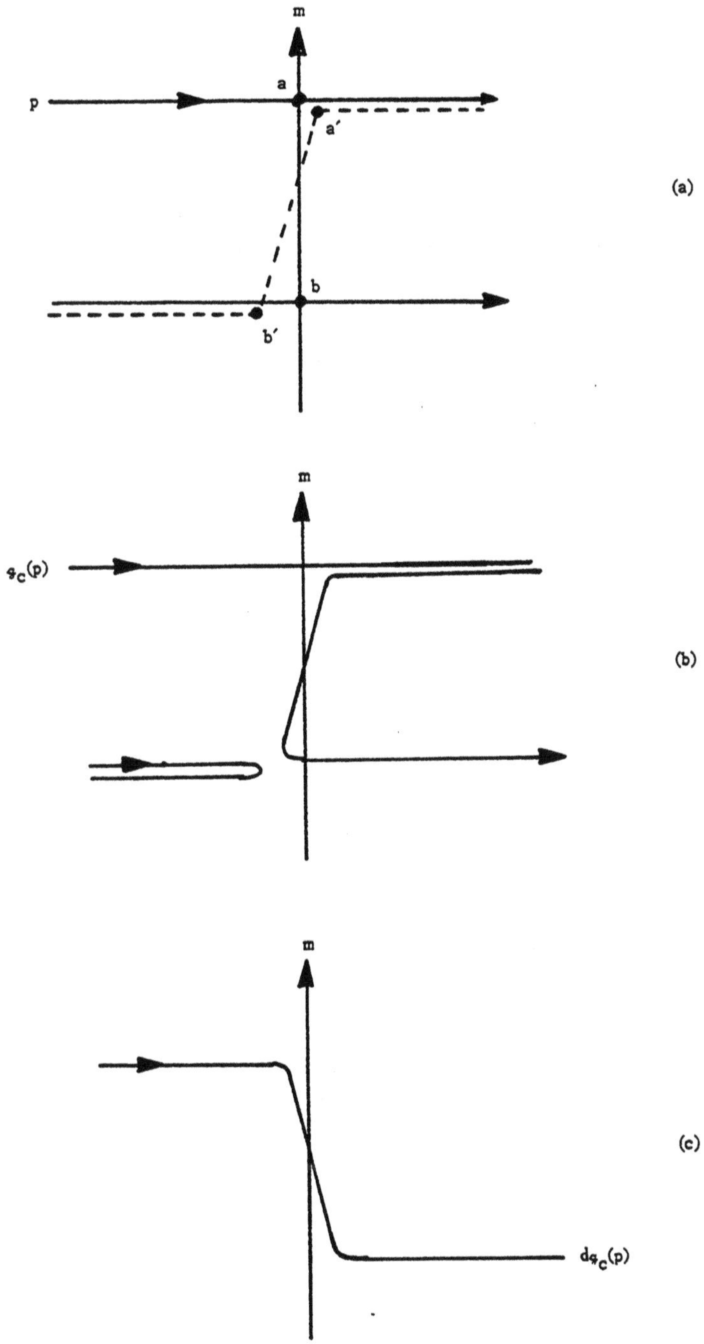

Fig. 19.

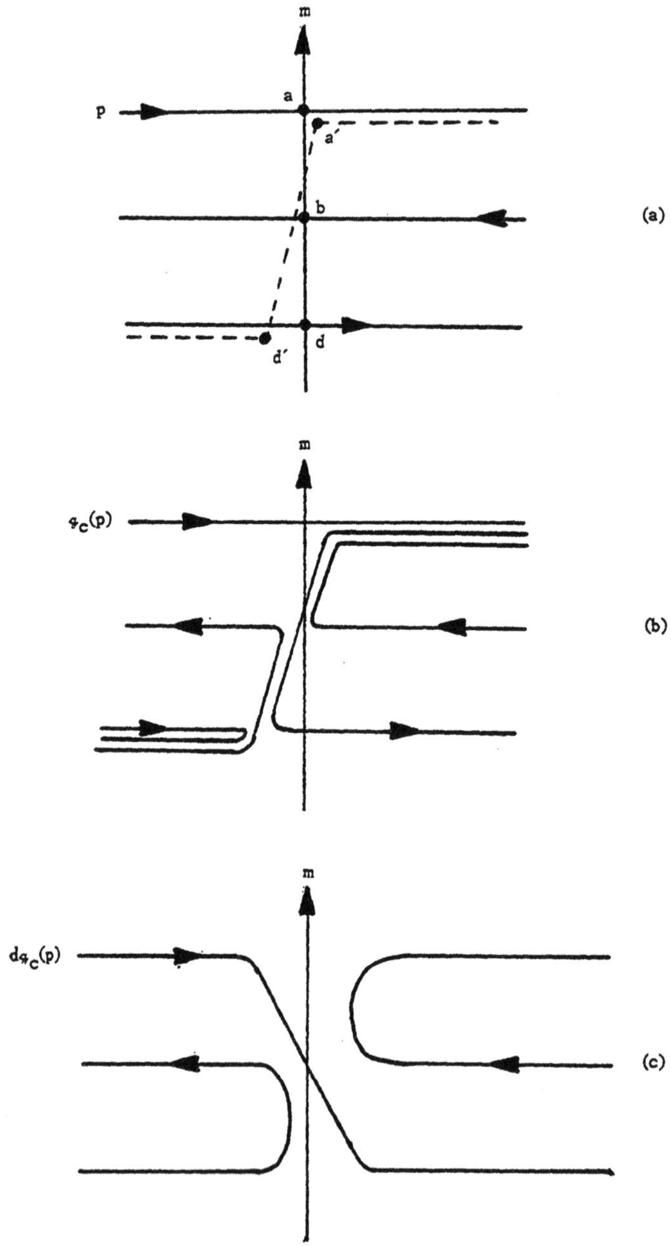

Fig. 20.

Case 2. Suppose that three consecutive crossing points a, b, d on m are such that p is oriented in alternating directions with respect to m at a, b and d respectively, as in Figure 20(a). One of the p-segments $\overrightarrow{ad}$ or $\overrightarrow{da}$ does not contain b; assume that the notation is chosen so that this segment is $\overrightarrow{da}$. Choose points d′; a′ in N, close to d, a respectively, and as illustrated in Figure 20(a) These points are to be chosen in such a way that they are both on the same side of p, with respect to the given orientation. Let c be a simple path which starts at a′ and proceeds, close to p, to the point d′, then crosses m once in N to return to a′. Figure 20(b) shows $\mathscr{g}_c(p)$. Then we may choose an element $d \in \mathfrak{D}$, such that $d\mathscr{g}_c(p) \subset p \cup N$, and also $|d\mathscr{g}_c(p) \cap m| < k$, as in Figure 20(c).[7] The induction hypothesis may now be applied to construct u. ||

LEMMA 4.6.4. *Let* $p, m_1, \cdots, m_r$ *be simple paths on* T_g, *with* p *being closed, and with* $|m_i \cap m_j| = 0$ *if* $i \neq j$, $1 \leq i, j \leq r$. *Then there exists a path* u *such that* $u \sim_c p$ *and such that for each* $i = 1, \cdots, r$ *either* $|u \cap m_i| = 0$ *or* 2_0.

Proof of Lemma 4.6.4. Since $|m_i \cap m_j| = 0$ we may find disjoint neighborhoods $N_1, \cdots, N_r$ of $m_1, \cdots, m_r$ respectively. Applying the procedure of Lemma 4.6.3 to $p_1 = p$ and m_1, we may find a path p_2 which satisfies the conditions of Lemma 4.6.4 with respect to the path m_1. Applying Lemma 4.6.3 a second time, to p_2 and m_2, we may find a path p_3 which satisfies the desired conditions with respect to m_2, and $|p_3 \cap m_1| = 0, 1$ or 2_0. Iterating this procedure r times, we obtain a path $u = p_r \sim_c p_{r-1} \sim_c \cdots \sim_c p_1 = p$ such that $|u \cap m_i| = 0, 1$ or 2_0 for each m_i. If, for some $i = 1, \cdots, r$, we have $|u \cap m_i| = 1$, then, by application of Lemma 4.6.2, we may obtain a path $u' \sim_c u$ such that $|u' \cap m_i| = 0$. ||

---

7 Note that $|c \cap m| < |p \cap m|$.

The techniques developed in Lemmas 4.6.1 - 4.6.4 will now be applied to examine a particular curve p on T_g and its intersections with the family of curves $\{m_i, a_i, d_i;\ 1 \le i \le g\}$ in Figure 21.

LEMMA 4.6.5. *Let* $\{m_i, d_i;\ 1 \le i \le g\}$ *be the simple closed curves on* T_g *which are illustrated in Figure 21. Let* $\mathfrak{h} \in \mathcal{G}$, *and let* $p = \mathfrak{h}(m_1)$. *Then there exists a simple closed curve* v *such that* $v \sim_c p$, $|v \cap m_i| = 0$ *and* $|v \cap d_i| = 0$ *or* 2_0, *for each* $i = 1, \cdots, g$.

Proof of Lemma 4.6.5. Note that the curves in our collection are pairwise disjoint. Hence, by Lemma 4.6.4, we may find a path $u \sim_c p$ such that $|u \cap m_i| = 0$ or 2_0 and $|u \cap d_i| = 0$ or 2_0 for each $i = 1, \cdots, g$. We will now show that $u \approx v$, where v does not meet any m_i.

If $|u \cap m_j| = 0$ for each $j = 1, \cdots, g$ we are done. Suppose, then, that for some j, $1 \le j \le g$, we had $|u \cap m_j| = 2_0$. Suppose also that $|u \cap d_j| = 0$. Note that d_i bounds a torus. Enumerating the various possibilities for the path u, one sees easily that u must either intersect itself (which is impossible) or bound a disc on T_g (which is likewise impossible because $u \sim_c p = \mathfrak{h}(m_1)$, and m_1 does not bound a disc). It then follows that if $|u \cap m_j| = 2_0$, then also $|u \cap d_j| = 2_0$. Let $u \cap m_j = \{a, b\}$. Then a and b divide u and m_j into pairs of arcs, $u = u_1 \cup u_2$ and $m_j = m_{j_1} \cup m_{j_2}$. At least one of the four pairs $u_1 \cup m_{j_1}$, $u_1 \cup m_{j_2}$, $u_2 \cup m_{j_1}$, $u_2 \cup m_{j_2}$ must bound a disc D_j in the jth handle, and using this disc we may construct a homeomorphism $d_j \in \mathfrak{D}$, with d_j supported in a neighborhood of D_j, such that $|d_j(u) \cap m_j| = 0$. A similar argument applied to each handle in turn yields the desired path $v = d_g \cdots d_2 d_1(u)$, where we choose $d_j = \mathrm{id}$ whenever $|u \cap m_j| = 0$. ||

LEMMA 4.6.6. *Let* p, m_1 *be as in Lemma 4.6.5. Then* $p \sim_c m_1$.

Proof of Lemma 4.6.6. Let $v \sim_c p$ be the simple closed curve whose existence is guaranteed by Lemma 4.6.5. Since $|v \cap m_j| = 0$ for

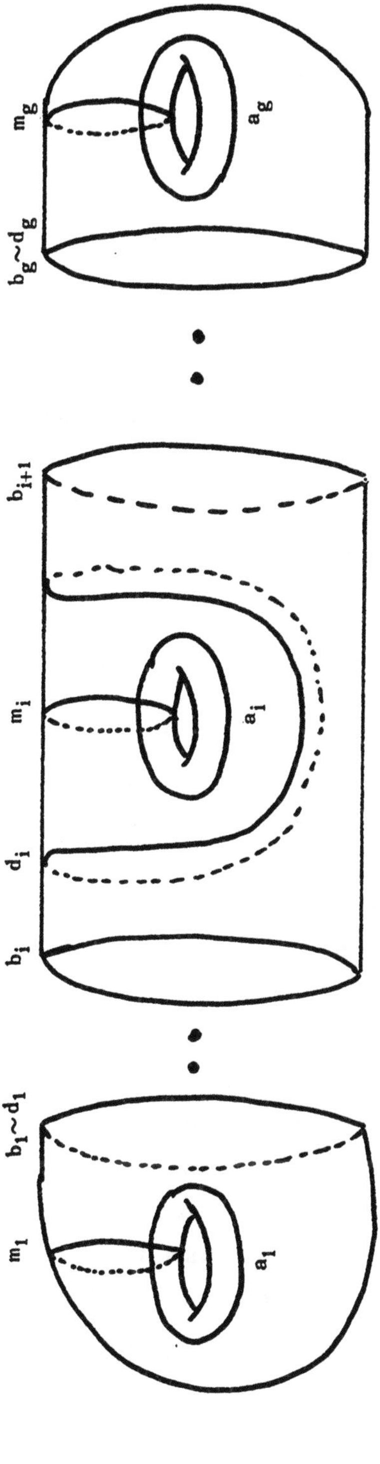

Fig. 21.

$j = 1, \cdots, g$, therefore $v \in T_g - m_1 \cup m_2 \cup \cdots \cup m_g$, i.e., a sphere with 2g discs removed. Now, v must divide the sphere (every simple closed curve does) but v does not divide T_g (because v is the homeomorphic image of m_1, and m_1 does not divide T_g). Therefore v must enclose one of the "boundary curves" of the sphere, hence v must meet one of the curves a_k $(1 \leq k \leq g)$ in Figure 21. Note that (by Lemma 4.6.5 we have $|v \cap m_k| = 0$ and $|v \cap d_k| = 0$ or 2_0. It then follows that the only possibility (since $|v \cap a_k| \neq 0$) is $|v \cap a_k|$ is odd, is that there is $v' \approx v$ such that $|v' \cap a_k| = 1$. Hence, by Lemma 4.6.2, we have $v \sim_c a_k$. Let m be a simple closed curve on T_g which meets a_k and m_1 once (if $k \neq 1$), or if $k \doteq 1$, let $m = m_1$. Then we may apply Lemma 4.6.2 again to obtain $a_k \sim_c m \sim_c m_1$. Since $\sim_c$ is a equivalence relation, and since $v \sim_c a_k$ and $p \sim_c v$, this implies $p \sim_c m_1$. ||

Proof of Theorem 4.6, part (i). Let $h \in \mathcal{G}$ be an arbitrary homeomorphism of T_g, and let $p = h(m_1)$. Then, by Lemma 4.6.6, we may find elements $y \in \mathcal{Y}$ and $d \in \mathfrak{D}$ such that $dy(p) = m_1$.

Assign an orientation to m_1. This induces an orientation on $dy(p) = m_1$, which may or may not agree with the orientation assigned to m_1. But now we observe (by a double application of Lemma 4.6.2) that the twist product $\vartheta_{a_1}\vartheta_{m_1}^2\vartheta_{a_1}$ maps m_1 to a curve which is isotopic to m_1 but with reversed orientation, so that, if necessary, we may replace dy with $d'y' = d'(\vartheta_{a_1}\vartheta_{m_1}^2\vartheta_{a_1})y$, $d' \in \mathfrak{D}$, in order to insure that the image of p has the same orientation as m_1. Let $h_* = dyh$ or $d'y'h'$, whichever is required. Then, by composing h_* with an appropriate isotopy, if necessary, we may assume that h_* restricted to m_1 is the identity.

Cutting open the surface T_g along the curve m_1, we may now regard h_* as a self-homeomorphism of T_g minus two discs, D_1 and D_2, where h_* is the identity on ∂D_i, $i = 1, 2$, because h_* is orientation-preserving in T_g. Let z_1^0 and z_2^0 be interior points of D_1, D_2 respectively. Then h_* may be extended from a map of $(T_{g-1} - D_1 \cup D_2) \to (T_{g-1} - D_1 \cup D_2)$ to a homeomorphism $\bar{h}_*: T_{g-1} \to T_{g-1}$ which keeps the points z_1^0, z_2^0 fixed.

Moreover, $\mathcal{h}_*$ will be isotopic to the identity map if and only if $\bar{\mathcal{h}}_*$ is isotopic to the identity map. We may interpret the isotopy class of $\bar{\mathcal{h}}_*$ as an element of the group $\pi_0 \mathcal{F}_2 T_{g-1}$.

Induction on g is now in order. The group M(0, 0) is, by Theorem 4.4, generated by twists (trivially). Assume, inductively, that M(g−1, 0) is generated by the isotopy classes of Dehn twists. Let $i_*: \pi_0 \mathcal{F}_2 T_{g-1} \to \pi_0 \mathcal{F}_0 T_{g-1} = M(g-1, 0)$ be the homomorphism defined in Section 4.1 and studied in Theorem 4.2. Then $\pi_0 \mathcal{F}_2 T_{g-1}$ is generated by ker i_* and the lifts of the generators of M(g−1, 0) to $\pi_0 \mathcal{F}_2 T_{g-1}$. By Lemma 4.6.1, every Dehn twist on $T_{g-1,0}$ is isotopic to a Dehn twist which lifts to a Dehn twist on $T_{g-1,2}$ (if the curve we are twisting about includes the point z_1^0 or z_2^0, we just move it a little bit). By Theorem 4.2, and the discussion which follows it at the end of Section 4.1, ker i_* is generated by "spins" of z_1^0 and z_2^0 about appropriate curves. But now observe that a spin of z_i^0 about c is just a product of Dehn twists, in opposite directions, about curves c_1 and c_2 such that $c_1 \approx c \approx c_2$, but c_1 and c_2 are separated by $z_i^0 \in c$. Thus $\bar{\mathcal{h}}_*$ is isotopic to a product of Dehn twists, hence $\mathcal{h}$ is isotopic to a product of Dehn twists. This completes the proof of Theorem 4.6, part (i). ||

Outline of proof of Theorem 4.6, part (ii). Note that, from the result in part (i), it will be adequate to establish that if v is an arbitrary simple closed curve on T_g, then a Dehn twist about v is isotopic to a product of Dehn twists about the curves $c_1, \cdots, c_{3g-1}$ in Figure 17. The proof will require one additional fact, which was not needed in part (i):

LEMMA 4.6.7. *Let* p, m *be simple closed curves on* T_g, *and suppose that* $\mathcal{h}(p) = m$, $\mathcal{h} \in \mathcal{G}$. *Then*

$$\mathcal{g}_m = \mathcal{h} \mathcal{g}_p \mathcal{h}^{-1} . \tag{4-14}$$

Proof of Lemma 4.6.7. This is an immediate consequence of the definition of a Dehn twist. ||

We consider a set of reference paths on the surface T_g. This set will be denoted R, and will include the paths $\{m_i, a_i, d_i, b_i;\ 1 \le i \le g\}$ shown in Figure 21, minus a small neighborhood of the g points $\{a_i \cap m_i;\ 1 \le i \le g\}$. (This small neighborhood is deleted in order to obtain a reference set which contains only *disjoint* paths; this will be needed to apply Lemma 4.6.8.):

LEMMA 4.6.8. *Let* $v, p_1, \cdots, p_r$ *be simple paths on* T_g, *with* v *being closed, and with* $|p_i \cap p_j| = 0$ *if* $i \neq j,\ 1 \le i, j \le r$. *Then there exists a path* $u \sim_c v$ *such that for each* $i = 1, \cdots, r$ *either* $|u \cap p_i| = 0, 1$ *or* 2_0. *Moreover, the twists in the equivalence* $u \sim_c v$ *satisfy the condition* $|c \cap \bigcup_{i=1}^{r} p_i| < |v \cap \bigcup_{i=1}^{r} p_i|$.

Proof of Lemma 4.6.8. See the demonstration of Lemma 4.6.3 and Lemma 4.6.4 and also the footnotes on pp. 172 and 175.||

Let v be our arbitrary simple closed curve on T_g. Since we are only interested in Dehn twists about v *up to isotopy* we may (by Lemma 4.6.1) assume that v avoids the point set $\{a_i \cap m_i;\ 1 \le i \le g\}$. We now assume (inductively) that, for all paths c which meet R in fewer than r points and avoid the points $a_i \cap m_i$, the Dehn twist ϑ_c is isotopic to a power product of Dehn twists about $c_1, \cdots, c_{3g-1}$. Suppose now that v meets R in precisely r points, $r \ge 2$.

Since the paths in R are disjoint, we may apply Lemma 4.6.8 to produce a simple closed curve u such that $u \sim_c v$, and such that if p is any path in R, then $|u \cap p| = 0, 1$ or 2_0.

Let $\mathscr{y} \in \mathscr{Y}$ be the power product of twists which is required in the equivalence $v \sim_c u$. Then, by Lemma 4.6.7, it must be true that $\vartheta_v = \mathscr{y}\vartheta_u\mathscr{y}^{-1}$. Now, $\mathscr{y} \in \mathscr{Y}$, and moreover $\mathscr{y}$ is a power product of Dehn twists about curves c which, by Lemma 4.6.8, meet R *less* than r times. Also, ϑ_u is a Dehn twist about the curve u, and $|u \cap p| = 0, 1$ or 2_0, $p \in R$. Hence the theorem will be true *if it can be established for all possible* cases *for which* $|u \cap p| = 0, 1$ *or* 2_0, $p \in R$.

This last step of the proof is accomplished by a (painful) enumeration of possibilities, and we omit it. The reader is referred to [Lickorish, 1964 and 1966] for details. Note that [Lickorish, 1966] corrects an error in this part of the proof in [Lickorish, 1964], hence the two papers must be read simultaneously.

We remark that it should be possible to conclude the proof of part (ii) of Theorem 4.6 by an induction on genus, utilizing the results of Theorem 4.3, just as we did for part (i). ‖

4.4. *Lifting and projecting homeomorphisms*

The reader may recall that in Section 4.2 we defined a special type of homeomorphism of the n-punctured sphere onto itself which resulted in an interchange of the ith and $(i+1)^{st}$ punctures, q_i and q_{i+1}. The isotopy classes $\{\omega_i; i = 1, \cdots, n-1\}$ of these maps were shown to generate the mapping class group $M(0, n)$ of $(S^2 - q_1, \cdots, q_{n-1})$. It is easy to see that these maps could be interpreted as special cases of Dehn twists. In this section we will explore this relationship.

We wish to represent the surface $T_{g,0}$ as a branched covering space of the sphere.[8] Suppose that $T_{g,0}$ is embedded in E^3 in the manner illustrated in Figure 17, so that $T_{g,0}$ is invariant under reflections in the xz and xy-planes. Let $c_1, c_3, \cdots, c_{2g+1}$ denote the intersections of $T_{g,0}$ with the xy plane, and let $c_2, c_4, \cdots, c_{2g}$ denote the intersections of $T_{g,0}$ with the xz-plane. Each c_i is a circle. Let $i : E^3 \to E^3$ be defined by $i(x, y, z) = (x, -y, -z)$. Observe that i maps $T_{g,0} \to T_{g,0}$, and

---

[8]   If $\widetilde{X}$ and X are triangulated n-manifolds, and if $p : \widetilde{X} \to X$ is a simplicial map, then p is said to be a *branched covering space projection* if the restriction of p to the complement of the (n–2)-dimensional skeleton of the triangulation is a covering space projection. The *branch set* $B \subset X$ is the collection of points $x \in X$ which have the property that x has no neighborhood U such that the restriction of p to an arc-component of $p^{-1}(U)$ is a covering map. The set $p^{-1}(B) = \widetilde{B}$ is the *branch cover*.

has order 2. The orbit space of $T_{g,0}$ under the action of the involution i will be homeomorphic to a sphere, $T_{0,0}$; this action induces a natural mapping $p: T_{g,0} \to T_{0,0}$, which is easily seen to be a branched covering space projection. The covering will be 2-to-1, except for the $(2g+2)$-fixed points of i, which occur along the x axis and are denoted $\tilde{q}_1, \cdots, \tilde{q}_{2g+2}$ in Figure 17. The image of each $\tilde{q}_i$ under p will be denoted, as before, by $q_i \in T_{0,0} = T_{g,0}/i \approx S^2$. Thus, if $\tilde{Q} = \{\tilde{q}_1, \cdots, \tilde{q}_{2g+2}\}$ and $Q = \{q_1, \cdots, q_{2g+2}\}$, then $p: (T_{g,0}, \tilde{Q}) \to (T_{0,0}, Q)$ will be a branched covering space projection, with branch set Q and branch cover $\tilde{Q}$. The involution i generates the group of covering transformation, which is of order 2.

Let N_i be a cylindrical neighborhood of c_i on T_g, which is parametrized by cylindrical coordinates (y, θ), $-1 \le y \le +1$, with c_i the circle $y = 0$, and $\tilde{q}_{2i-1} = (0, 0)$, $\tilde{q}_{2i} = (0, \pi)$. Then the twist $\mathcal{T}_{c_i}$ is defined by equation (4-13). It will be assumed that the parametrization is such that $\mathcal{T}_{c_i}$ is fiber-preserving. The image of N_i under p will be a disc D_i. For odd i, $\partial D_i = p\{y=-1\} = p\{y=+1\}$, and $p(c_i)$ will be an arc joining $q_{2i-1}, q_{2i} \in D_i$. If $c'_i = p\{y=\frac{1}{2}\}$, then the projection $p\mathcal{T}_{c_i}p^{-1}$ of $\mathcal{T}_{c_i}$ is a twist $\mathcal{T}_{c'_i}$ about c'_i, which may be identified with the map h_i of Section 4.2. A similar argument holds if i is even. Thus $\mathcal{T}_{c_1}, \cdots, \mathcal{T}_{c_{2g+1}}$ (Theorem 4.6) project to representatives of $\omega_1, \cdots, \omega_{2g+1}$ (Theorem 4.5).

This does not, in itself, imply a relationship between the groups $M(g, 0)$ and $M(0, 2g+2)$, because elements in $M(0, 2g+2)$ are isotopy classes of maps with respect to isotopic deformations on the punctured sphere, while elements of $M(g, 0)$ are isotopy classes of maps with respect to isotopic deformations on the closed covering surface $T_{g,0}$. Surprisingly, it has been discovered that in the situation described above, and also for other appropriate covering spaces, entire classes of maps (with respect to isotopic deformations on the n-punctured sphere) lift to classes of maps, unique up to covering transformations, (with respect to isotopic deformations on the closed covering surface).

The development of these ideas, for k-fold cyclic coverings of the sphere and also for k-fold cyclic coverings of one Riemann surface by another, is given in three papers by Birman and Hilden [1971, 1972, and 1973]. New generalizations of these results have also been obtained by [Maclachlan and Harvey, 1973] and again by [Zieschang, 1973]. (The methods of proof used in the latter two papers are different from each other, and from the techniques described here; Maclachlan and Harvey, in particular, make an interesting observation in "Remark 2" following "Corollary 12" of their paper, indicating an appropriate generalization to n-dimensional manifolds.)

Our efforts here will be confined to a special case: we will adapt the methods of Birman and Hilden, 1973 to a special problem: The determination of defining relations for the group $M(2,0)$. We begin by establishing:

THEOREM 4.7. *Let* $g = 2$. *Let* $h : T_{2,0} \to T_{2,0}$ *be a homeomorphism which is fiber-preserving*[9] *with respect to the covering space projection* $p : (T_{2,0}, \tilde{Q}) \to (T_{0,0}, Q)$. *Suppose also that* h *is isotopic to the identity map. Then there is an isotopy* h_t *between* $h = h_0$ *and* $\mathrm{id} = h_1$ *such that for each* $0 \le t \le 1$ *the map* h_t *is fiber-preserving.*

We will then use Theorem 4.7 to establish:

THEOREM 4.8. *The mapping class group* $M(2,0)$ *of a closed, orientable surface of genus* 2 *admits the presentation:*

---

9 A homeomorphism h of $T_{2,0} \to T_{2,0}$ is fiber-preserving if, for all pairs $z, z' \in T_{2,0}$ such that $p(z) = p(z')$, it is also true that $ph(z) = ph(z')$. Note that h is fiber-preserving if and only if h commutes with the involution i.

generators: $\zeta_1,\cdots,\zeta_5$

definiting relations:

(4-15) $$\zeta_i\zeta_j = \zeta_j\zeta_i \quad \text{if } |i-j| \geq 2,\ 1 \leq i, j \leq 5$$

(4-16) $$\zeta_i\zeta_{i+1}\zeta_i = \zeta_{i+1}\zeta_i\zeta_{i+1} \quad (1 \leq i \leq 4)$$

(4-17) $$(\zeta_1\zeta_2\cdots\zeta_5)^6 = 1$$

(4-18) $$(\zeta_1\zeta_2\zeta_3\zeta_4\zeta_5^2\zeta_4\zeta_3\zeta_2\zeta_1)^2 = 1$$

(4-19) $$\zeta_1\zeta_2\zeta_3\zeta_4\zeta_5^2\zeta_4\zeta_3\zeta_2\zeta_1 \rightleftarrows \zeta_i, \quad 1 \leq i \leq 5\,.$$

The generator ζ_i $(1 \leq i \leq 5)$ *in this presentation may be interpreted geometrically as the isotopy class of a Dehn twist about the curve* c_i *in Figure 17 (taking* $g = 2$).

At the conclusion of this section, after the proofs of Theorems 4.7 and 4.8, we will discuss briefly the generalization of Theorems 4.7 and 4.8 to other covering space projections of one 2-manifold by another.

Proof of Theorem 4.7. Since h is a fiber-preserving map, it is immediate that h maps the set of exceptional points $\{\tilde{q}_1,\cdots,\tilde{q}_6\}$ onto itself, however it might permute the individual points in this set. We begin our proof by establishing that in fact the only possibility is that $h(\tilde{q}_i) = \tilde{q}_i$, $1 \leq i \leq 6$.

Suppose that $h(\tilde{q}_i) = \tilde{q}_j$ for some $i \neq j$. Let γ be any $\tilde{q}_i$-based loop on $T_{2,0}$. Since h preserves fibers,

(4-20) $$h\,i(\gamma) = i\,h(\gamma)\,.$$

Let h_t be the given isotopy between the homeomorphism h and the identity map. Let β denote the path $\beta(t) = h_t(\tilde{q}_i)$ joining $\tilde{q}_i$ to $\tilde{q}_j$. Then

(4-21) $$\gamma \cong \beta h(\gamma)\beta^{-1}$$

where the homotopy taking γ to $\beta h(\gamma)\beta^{-1}$ is defined by the isotopy $h_s(\gamma(t))$. Applying i to the homotopy in (4-21) and using (4-20) we obtain

(4-22) $$i(\gamma) \cong (i(\beta))(h(i(\gamma)))(i(\beta))^{-1}.$$

Now consider the $\bar{q}_i$-based loop $\iota(\gamma)$. Just as in (4-21), we have

$$\iota(\gamma) \cong \beta \hbar(\iota(\gamma))\beta^{-1} . \tag{4-23}$$

Combining (4-22) and (4-23), we see that $\beta^{-1}\iota(\beta)$, which is a closed loop based on $\bar{q}_j$, commutes with $\hbar(\iota(\gamma))$. Since γ was arbitrary, and $\hbar$ and ι are homeomorphisms, it follows that $\beta^{-1}\iota(\beta)$ commutes with every element of $\pi_1(T_{2,0}, \bar{q}_j)$. But the center $\pi_1(T_{2,0}, \bar{q}_j)$ is trivial [see Magnus, Karass and Solitar, 1966, Corollary 4.5], hence $\beta^{-1}\iota(\beta)$ must be homotopic to the constant loop, hence

$$\beta \cong \iota(\beta) \tag{4-24}$$

where the homotopy is a homotopy of paths joining $\bar{q}_i$ to $\bar{q}_j$, keeping endpoints fixed.

Let U be the universal covering surface of $T_{2,0}$. Then U is hyperbolic, and ι lifts to a Moebius transformation $\hat{\iota}$ of U. By composing with a covering transformation if necessary, we may assume that $\hat{\iota}(\hat{q}_j) = \hat{q}_j$ for some $\hat{q}_j$ lying over $\bar{q}_j$. Since $\beta \cong \iota(\beta)$, we also have $\hat{\iota}(\hat{q}_i) = \hat{q}_i$, where $\hat{q}_i$ is the unique endpoint of the lift of β^{-1} which begins at $\hat{q}_j$. Since $\bar{q}_i \neq \bar{q}_j$, it follows that $\hat{q}_i \neq \hat{q}_j$. But then $\hat{\iota}$ is a Moebius transformation which has two fixed points. This is impossible, hence $\hat{q}_i = \hat{q}_j$, hence $\bar{q}_i = \bar{q}_j$.

We now examine the $\bar{q}_i$-based loop $\beta(t) = \hbar_t(\bar{q}_i)$, and see that in fact it must be homotopic to the constant loop in $\pi_1(T_{2,0}, \bar{q}_i)$. To see this, observe that the previous argument implies that the closed $\bar{q}_i$-based loop $\beta(t)$ on $T_{2,0}$ lifts to a closed $\hat{q}_i$-based loop on U. Since U is simply-connected, the lift of $\beta(t)$ is homotopic to the identity. Projecting the homotopy to $T_{2,0}$ we obtain:

$$\beta \cong 1 . \tag{4-25}$$

To proceed further, we will need a lemma:

LEMMA 4.7.1. *Let* q *be a point in a* p.l. *manifold* X *without boundary. Let* $\beta(t)$ *be a* q*-based loop in* X *which is homotopic to the constant loop at* q. *Then there is an isotopy* ℓ_t *of* X *such that* $\ell_0 = \ell_1 = \mathrm{id}$, *where* ℓ_t *has compact support, and* $\ell_t(q) = \beta(t)$.

Proof. The proof of this lemma follows from the simplicial approximation theorem and the 2-isotopy extension theorem [see p. 154 of Hudson, 1969]. ||

We can now continue the proof of Theorem 4.7, by using the isotopy h_t between h and the identity map to construct a new isotopy h'_t between h and the identity, where h'_t will have the additional property that $h'_t(\tilde{q}_i) = \tilde{q}_i$ for every $i = 1, \cdots, 6$ and for every $t \in [0,1]$. To construct h'_t, first observe that $h(\tilde{q}_1) = \tilde{q}_1$ and $h_t(q_1) = \beta_1(t) \cong 1$ in $\pi_1(T_{2,0}, \tilde{q}_1)$. By Lemma 4.7.1 there is an isotopy ℓ_t of $T_{2,0}$ with $\ell_0 = \ell_1 = \mathrm{id}$ and $\ell_t(\tilde{q}_1) = \beta_1(t)$. Let $h_t^{(1)} = \ell_t^{-1} h_t$. Then $h_t^{(1)}(\tilde{q}_1) = \tilde{q}_1$ for all $t \in [0,1]$, and $h_t^{(1)}$ is again an isotopy between h and id.

Let $T_{2,1} = T_{2,0} - \tilde{q}_1$; let $T_{0,1} = T_{0,0} - q_1$; let $p^{(1)} = p|T_{2,1}$; let $h^{(1)} = h|T_{2,1}$; and let $h_t^{(1)} = h_t|T_{2,1}$. Then we may repeat the entire argument from the beginning of the proof of Theorem 4.7, replacing $T_{2,0}$, $T_{0,0}$, p, h, h_t by $T_{2,1}$, $T_{0,1}$, $p^{(1)}$, $h^{(1)}$, $h_t^{(1)}$ respectively. After six such repetitions we will obtain an isotopy h'_t between h and the identity map which has the desired property that $h'_t(\tilde{q}_i) = \tilde{q}_i$ for each $i = 1, \cdots, 6$ and for each $t \in [0,1]$.

Summarizing the argument so far: we have shown that our fiber-preserving homeomorphism $h: T_{2,0} \to T_{2,0}$ may always be isotopied to the identity map via an isotopy h'_t which keeps each point $\tilde{q}_i$ fixed throughout the isotopy. But then, there is no loss in generality in assuming that the restriction of h to $T_{2,0} - \{\tilde{q}_1, \cdots, \tilde{q}_6\}$ is a fiber-preserving homeomorphism with respect to the unique *unbranched* covering space projection $\bar{p}: (T_{2,0} - \tilde{Q}) \to (T_{0,0} - Q)$, which is associated with our branched covering space $p: (T_{2,0}, \tilde{Q}) \to (T_{0,0}, Q)$ (the projection $\bar{p}$ is the restriction of p to $T_{2,0} - \tilde{Q}$). It remains to prove that the isotopy h'_t can, in turn, be replaced by a new isotopy h''_t which is fiber-preserving for every $t \in [0,1]$.

Since h is fiber-preserving, it projects to $h^{\#}: T_{0,6} \to T_{0,6}$. Choose points $z \in T_{2,6}$, $z^{\#} \in T_{0,6}$ such that $p(z) = z^{\#}$. Let P denote the curve $h'_t(z)$, and let $\beta^{\#}$ be the projection of β. We may define $h^{\#}_*$, an automorphism of $\pi_1(T_{0,6}, z^{\#})$ by

$$\text{(4-26)} \qquad h^{\#}_*[\gamma^{\#}] = (\beta^{\#})(h^{\#}(\gamma^{\#}))(\beta^{\#})^{-1} \, ,$$

for each $[\gamma^{\#}] \in \pi_1(T_{0,6}, z^{\#})$. Let γ be a z-based loop, and let $\gamma^{\#}$ be the projection of γ. Then $[\gamma^{\#}] \in \bar{p}_* \pi_1(T_{2,6}, z)$, and we have $h^{\#}_*[\gamma^{\#}] = [\gamma^{\#}]$, because $\gamma \cong \beta h(\gamma) \beta^{-1}$. (The homotopy is defined by the isotopy h'_t.) Thus we may assume that the restriction of $h^{\#}_*$ to $\bar{p}_* \pi_1(T_{2,6}, z)$ is the identity automorphism.

Now choose any element $[\alpha^{\#}] \in \pi_1(T_{2,6}, z)$ and any element $[\beta^{\#}] \in \pi_1(T_{0,6}, z^{\#})$. Since the covering is regular, it must be true that $[\beta^{\#} \alpha^{\#} (\beta^{\#})^{-1}] \in \bar{p}_* \pi_1(T_{2,6}, z)$. Thus

$$\text{(4-27)} \qquad [\beta^{\#} \alpha^{\#} (\beta^{\#})^{-1}] = h^{\#}_*[\beta^{\#} \alpha^{\#} (\beta^{\#})^{-1}] = (h^{\#}_*[\beta^{\#}]) [\alpha^{\#}] (h^{\#}_*[\beta^{\#}])^{-1}.$$

Since $\alpha^{\#}$ was arbitrary, it follows that $[\beta^{\#}]^{-1} (h^{\#}_*[\beta^{\#}])$ must be in the centralizer of the group $\bar{p}_* \pi_1(T_{2,6}, z)$ in $\pi_1(T_{0,6}, z^{\#})$. But $\pi_1(T_{0,6}, z^{\#})$ is a free group of rank 5, hence all of its subgroups are free, hence any non-cyclic subgroup has a trivial centralizer. Thus $\bar{p}_* \pi_1(T_{2,6}, z)$ has a trivial centralizer, because it is of finite index and therefore is non-cyclic. Therefore we must have

$$\text{(4-28)} \qquad \beta^{\#} = h^{\#}_*(\beta^{\#}) \, .$$

Since $T_{0,6}$ is a surface, this implies [see Mangler, 1939] that $h^{\#}$ is isotopic to the identity, via an isotopy $h^{\#}_t$. Let h''_t be the lift of $h^{\#}_t$ to $T_{2,6}$. Then h''_t is a fiber-isotopy taking h to the identity. This completes the proof of Theorem 4.7. ‖

Proof of Theorem 4.8. The reader is reminded that in Theorem 4.5 we obtained a presentation for the group $M(0, n)$, which we may now specialize to the case $n = 6$. The generators of $M(0, 6)$ are thus

$\omega_1, \cdots, \omega_5$, each ω_i being the isotopy class of a map h_i on $T_{0,6}$ which interchanges the points q_i and q_{i+1}, as defined on p. 165. Since h_i lifts (modulo i) to $T_{2,6}$, and since isotopies of h_i lift to fiber-isotopies of $T_{2,6}$, it follows that there is a natural homomorphism from $M(0,6)$ to $M(2,6)/Z_2$ where Z_2 is the cyclic subgroup of order 2 generated by the isotopy class $[i]$ of i. It is clear that this homomorphism induces a homomorphism from $M(0,6)$ to $M(2,0)/Z_2$. The latter homomorphism is onto, because the isotopy classes of the twist maps $g_{c_i} (i = 1, \cdots, 5)$, which are the lifts of the map h_i on $T_{0,6}$, generate the group $M(2,0)$, by Theorem 4.6, part (ii). The homomorphism is invertible, by Theorem 4.7 and because lifting and projecting fiber-preserving homeomorphisms are mutually inverse operations. Hence $M(0,6)$ is isomorphic to $M(2,0)/Z_2$, under the mapping $\omega_i \to \zeta_i$, $1 \le i \le 5$.

Since $[i]$ has order 2 and is in the center of $M(2,0)$, because i commutes with $\{g_{c_i}; i = 1, \cdots, 5\}$, we are able to use the isomorphism established above to obtain a presentation for $M(2,0)$ in terms of the generators $\zeta_1, \cdots, \zeta_5, [i]$. Defining relations will include:

(i)   The relation $[i]^2 = 1$.

(ii) The relations which express the fact that $[i]$ commutes with $\zeta_1, \cdots, \zeta_5$.

(iii) The lifts of relations (4-9)-(4-12) for the case $n = 6$, to the group $M(2,0)$. One verifies without difficulty that relations (4-9), (4-10) and (4-12) lift to relations (4-15), (4-16) and (4-17) respectively. Relation (4-11) lifts to:

$$(4\text{-}29) \qquad \zeta_1 \zeta_2 \zeta_3 \zeta_4 \zeta_5^2 \zeta_4 \zeta_3 \zeta_2 \zeta_1 = [i] \ .$$

This fact, combined with the observation in (i) and (ii) above, gives relations (4-18) and (4-19). Thus (4-15)−(4-19) are a complete set of defining relations in $M(2,0)$. ‖

One might hope that the methods of Theorems 4.7 and 4.8 could be generalized to surfaces of arbitrary genus g, and indeed they can. However in the general case the subgroup of M(g, 0) which is obtained by lifting and projecting homeomorphisms is a proper subgroup of M(g, 0). This is easily seen to be the case for 2-sheeted coverings of the sphere, by observing that the simple twists which represent $\omega_1, \cdots, \omega_{2g+2}$ lift to Dehn twists which represent the generators $\zeta_1, \cdots, \zeta_{2g+2}$ of M(g, 0), but from Theorem 4.6 the group M(g, 0) is generated by $\zeta_1, \cdots, \zeta_{2g+2}$, $\zeta_{2g+3}, \cdots, \zeta_{3g-1}$ if $g \geq 3$. For a full discussion of these ideas, the reader is referred to [Birman and Hilden, 1973].

4.5. *Survey: mapping class groups of 2-manifolds*

The mapping class groups M(g, n) have potentially important application in several areas of mathematics, so that one would like to be able to learn more about these groups. For example, one approach to the study of 3-manifolds is to construct 3-manifolds from two handlebodies X_g, X'_g, which are "sewn" together along their boundaries by means of a surface homeomorphisms which identifies points on ∂X_g and $\partial X'_g$ [see Birman, 1973b; also Viro, 1972; Birman and Hilden, 1973a and 1974]. Clearly each such homeomorphism can be identified with an element of the mapping class group M(g, 0). The chief obstacle to such an approach has been the lack of knowledge about the groups M(g, 0). Equally important applications occur in Teichmuller theory, and again in infinite group theory. We close this monograph with a brief summary of the present state of knowledge of these groups.

1. The group M(g, n) maps homomorphically onto the groups M(g, 0). The kernel of this homomorphism was identified in Theorem 4.3, where it was shown to be isomorphic to the n-string braid group of the surface, except for a handful of special cases. The braid groups are in principle computable, by the methods described in Chapter 1. Moreover, their structural properties are fairly well-understood. Hence the groups M(g, n) should be accessible, if M(g, 0) were accessible.

2. The groups M(g, 0) are well known for $g = 0$ and 1. A presentation for M(2, 0) is given in Theorem 4.8. One expects that this result

will lead to other new results, for example, about Heegaard splittings of genus 2.

3. If $g \geq 3$, a finite set of generators are known for $M(g, 0)$ (Theorem 4.6) but defining relations are not known. The methods used to establish Theorems 4.7 and 4.8 can also be used to obtain presentations for large subgroups of $M(g, 0)$, but it has been proved that none of these can coincide with $M(g, 0)$ if $g \geq 3$ [see Maclachlan and Harvey, 1973]. The subgroups so obtained may be described algebraically as the normalizers of all subgroups of finite order in $M(g, 0)$. It is an open question whether *every* element in $M(g, 0)$ is in the normalizer of *some* element of finite order. (Our conjecture is a weak "no".) It is, however, easy to see that the group $M(g, 0)$ is *generated* by the centralizers of a particular set of $g-1$ involutions on the surface $T_{g,0}$.

4. The mapping class groups of non-orientable surfaces are closely related to those of orientable surfaces. This relationship arises by regarding the orientable surface as a double covering of the non-orientable surface. For details, see Birman and Chillingworth, 1972. Thus most questions about mapping class groups of non-orientable surfaces can be rephrased as questions about the mapping class group of an orientable double covering. Again, the groups $M(g, 0)$, $g \geq 3$, are the chief obstacle to further progress.

5. The group $M(g, 0)$ has a natural homomorphic image in the group of $2g \times 2g$ "symplectic" matrices with integral entries, $Sp(2g, Z)$. The latter group may be identified as the induced group of automorphisms of the abelianized group $(\pi_1 T_{g,0})/(\pi_1 T_{g,0})'$. This group is infinite, has a far-from-transparent structure, yet it is considerably more tractable than $M(g, 0)$, and many interesting questions can be posed about $Sp(2g, Z)$ and the kernel of the natural homomorphism from $M(g, 0)$ to $Sp(2g, Z)$. For a start, the interested reader is referred to [Hua and Reiner, 1949; Klingen, 1961; Birman, 1971].

6. By a recent result of E. Grossman, the groups $M(g, 0)$ are residually finite for every g. It would be interesting to identify finite homomorphic images of $M(g, 0)$ which do not factor through $Sp(2g, Z)$. A method

for constructing such representations has been discovered by R. Gillman, and will be reported on in the near future. The study of such representations will hopefully yield new insights into the structure of the mapping class groups.

CHAPTER 5

PLATS AND LINKS

Our emphasis in Chapters 1-4 has been on the relationship between closed braids and links. It would, however, be amiss to end this monograph without mention of another, and quite different, algebraic-topological connection between Artin's braid group and links in S^3, the concept of a "plat." This latter notion has only recently begun to receive attention, yet it appears to be of considerable importance in connection with the classification of 3-manifolds [see Viro, 1972; Birman and Hilden, 1973a and 1974; Montesinos, 1974], and it may also offer the possibility of a new approach to the study of knots and links in S^3.

Section 5.1 below is concerned with the definition of a plat, and the connections between plat representations of links and other representations. Section 5.2 contains a discussion of the "algebraic plat problem," that is, of the possibility of expressing the problem of equivalence of two links in S^3 as an algebraic problem about their plat representatives. The latter problem appears to be related to (or equivalent to) the problem of finding double coset representatives for a certain subgroup of the mapping class group $M(0,2m)$ of a sphere with an even number of points removed. In Section 5.3 we use the results of Sections 5.1 and 5.2 to give a constructive solution to the link problem for links with 2 bridges.

5.1. *Representing a link by a plat*

Let D^3 be a 3-ball of radius 2 centered at the origin in oriented 3-space E^3. Let $\mathfrak{A} = \mathfrak{A}_1 \cup \cdots \cup \mathfrak{A}_m$ be a collection of m unknotted and unlinked arcs which are embedded in D^3, with $Q_{2m} = \partial\mathfrak{A}$ a set of 2m points on $S^2 = \partial D^3$ (see Figure 22a). Let $t: E^3 \to E^3$ be a translation,

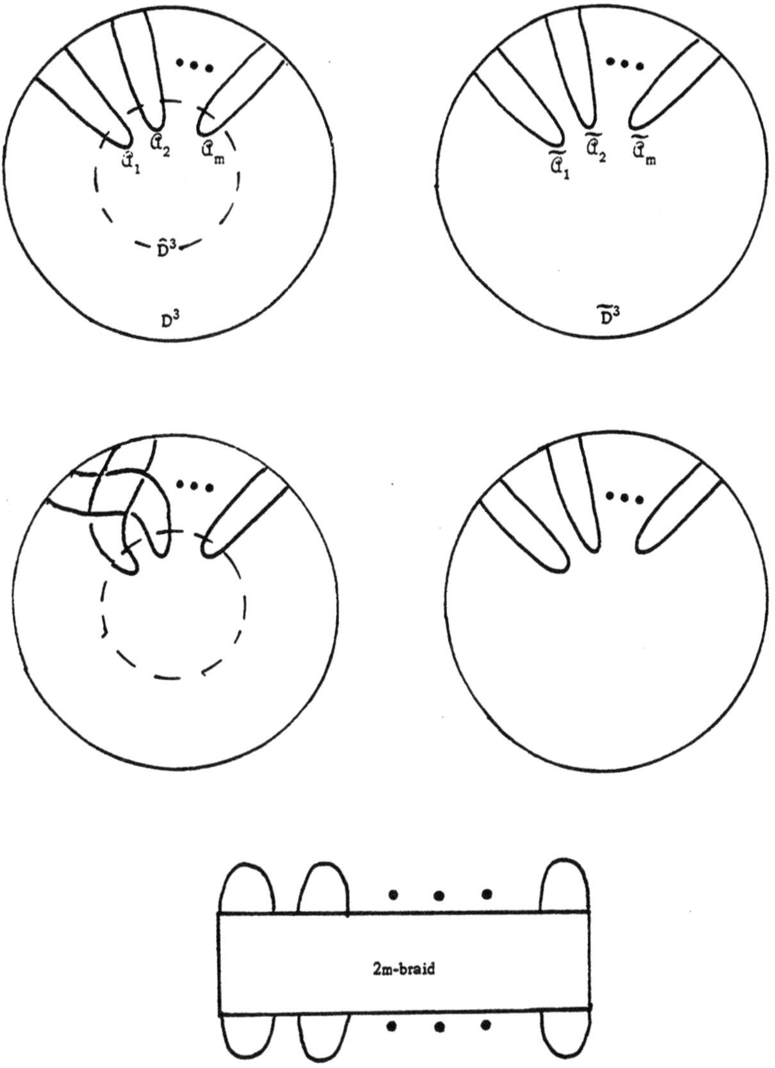

(a)

(b)

(c)

Fig. 22

and let $\tilde{D}^3 = t(D^3)$, $\tilde{\mathfrak{A}} = t(\mathfrak{A})$. Suppose, now, that $f:(S^2, Q_{2m}) \to (S^2, Q_{2m})$ is any orientation-preserving homeomorphism which keeps the set Q_{2m} fixed as a set. We may use the maps t and f to define an identification space $(D^3, \mathfrak{A}) \cup_f (\tilde{D}^3, \tilde{\mathfrak{A}})$ by pasting S^2 to S^2, using the rule $tf(p) = p$

for each point p on S^2. The space $D^3 \cup_f \tilde{D}^3$ is easily seen to be S^3, represented by a genus 0 Heegaard splitting [for the definition of a Heegaard splitting see, for example, Seifert and Trelfalls, 1934]. The subset $V = \mathfrak{A} \cup_f \tilde{\mathfrak{A}}$ will be a collection of $\mu \leq m$ disjoint simple closed curves in $S^3 = D^3 \cup_f \tilde{D}^3$, that is, a link. If a link is represented in this wav, then it is said to be displayed as a *plat on* $2m$ *strings*.

Recall that two links V, V^* are equivalent if there is a homeomorphism $h: S^3 \to S^3$ such that $h(V) = V^*$. Note that any isotopic deformation of the surface mapping f which leaves the point set Q_{2m} invariant can be extended to an isotopy of $(D^3, \mathfrak{A}) \cup_f (\tilde{D}^3, \tilde{\mathfrak{A}})$. It then follows that if f and f^* are isotopic maps of (S^2, Q_{2m}), then f and f^* will define equivalent plats $V = \mathfrak{A} \cup_f \tilde{\mathfrak{A}}$ and $V^* = \mathfrak{A} \cup_{f^*} \tilde{\mathfrak{A}}$. Hence our construction associates a link type with each element in the mapping class group $M(0, 2m)$ of the surface $S^2 - Q_{2m}$ (cf. Section 4.2). Of course, it is to be expected that distinct elements of $M(0, 2m)$ may define the same link type; this matter will be discussed in Section 5.2 below.

In the description above, it may be difficult for the reader to visualize the link $V = \mathfrak{A} \cup_f \tilde{\mathfrak{A}}$, because the arcs $\mathfrak{A} = \mathfrak{A}_1 \cup \cdots \cup \mathfrak{A}_m$ and $\tilde{\mathfrak{A}} = \tilde{\mathfrak{A}}_1 \cup \cdots \cup \tilde{\mathfrak{A}}_m$ are unknotted and unlinked, and the essential features of the link are concealed in the surface mapping f which is used to identify $\partial\mathfrak{A}$ with $\partial\tilde{\mathfrak{A}}$. We may, however, easily change the picture. Let $\hat{D}^3 \subset D^3$ be a solid ball of radius $r < 2$ concentric to D^3 which meets the set $\mathfrak{A}$ in a collection of m non-degenerate arcs $\hat{\mathfrak{A}}_1 \cup \cdots \cup \hat{\mathfrak{A}}_m$, with $\partial\hat{D}^3 \cap \mathfrak{A}$ being $2m$ distinct points (see Figure 22a). Note that $D^3 = \hat{D}^3 \cup (S^2 \times I)$, where $S^2 \times \{1\}$ is identified with $\partial\hat{D}^3$ and $S^2 \times \{0\}$ is identified with ∂D^3. Since $f: \partial D^3 \to \partial D^3$, it follows from Theorem 4.4 that f is isotopic to the identity map (via an isotopy which will in general move the $2m$ points in $\partial\mathfrak{A}$). Let f_t be the isotopy, with $f_0 = f$ and $f_1 =$ identity. We use the isotopy f_t to define a self-homeomorphism h of $S^3 = D^3 \cup_f D^3$ by the rule that h restricted to $\hat{D}^3$ and h restricted to $\tilde{D}^3$ are each the identity map, while the restriction of h to $S^2 \times I$ is defined by

(5-1) $$h(p,t) = (f_t(p),t), \qquad (p,t) \in S^2 \times I .$$

One may verify easily that h is well-defined on $\partial\hat{D}^3$ and on $\partial D^3 = \partial\tilde{D}^3$, hence $h(V) = h(\mathfrak{A}) \cup_{id} h(\tilde{\mathfrak{A}})$ represents the same link type as $V = \mathfrak{A} \cup_f \tilde{\mathfrak{A}}$. Figure 22b illustrates the link $h(V)$ in a typical case, and we see that it now looks like a link! It is, moreover, now clear that every plat may be constructed from a "geometric braid," as defined in Chapter 1 (cf. Figure 1) by identifying the strings in pairs, with the top (respectively bottom) of the $(2i-1)^{st}$ string identified with the top (respectively bottom) of the $2i^{th}$ string for each $i = 1,\cdots,m$, as in Figure 22c.

A natural question to ask is whether all links admit such a representation?

THEOREM 5.1. *Every link type may be represented (non-uniquely) as a plat.*

Proof. By Theorem 2.1, every link type may be represented (non-uniquely) as a closed braid, say on m strings. If a link type $\mathfrak{L}$ is represented as a closed braid obtained from a geometric m-braid β, then $\mathfrak{L}$ is also represented by the 2m-plat which is associated with the geometric 2m-braid $\beta_0\hat{\beta}\beta_0^{-1}$, where β_0 is a 2m braid which is defined in terms of the standard generators of the braid group by $\beta_0 = (\sigma_2\sigma_3\cdots\sigma_{2m-1})(\sigma_3\sigma_4\cdots\sigma_{2m-2})\cdots(\sigma_m\sigma_{m+1})$, and $\hat{\beta}$ is the 2m braid obtained from β by the addition of m trivially-braided strings. ||

By analogy with Theorem 2.2 of Chapter 2, we may now establish a new connection between braid automorphisms and link groups. As in earlier chapters, B_{2m} denotes the 2m-string Artin braid group, which (by Corollary 1.8.3) may be regarded as a group of automorphisms of the free group F_{2m} with free basis $x_1,\cdots,x_{2m}$. The reader is reminded that in Theorem 1.10 we interpreted the free group F_{2m} geometrically as the fundamental group of a disc D^2, with a set Q_{2m} of $2m$ points removed.

This allowed us to interpret B_{2m} as the mapping class group of $D^2 - Q_{2m}$, where admissible maps were required to keep ∂D^2 fixed pointwise. This theme was developed further in Chapter 4, where it was shown that the natural embedding of $(D^2 - Q_{2m})$ in $(S^2 - Q_{2m})$ induced by the inclusion of D^2 as a subset of S^2 induces a homomorphism from B_{2m} *onto* the mapping class group $M(0,2m)$ of $(S^2 - Q_{2m})$. This last assertion is the essential feature of the following result:

COROLLARY 5.1.1. *Let* $\beta \in B_{2m}$, *and suppose that the action of* β *on the free group* F_{2m} *is given by equation* (1-22), *with* $n = 2m$. *Let* $\mathfrak{B}$ *be the link type which is represented by the* $2m$ *plat defined by* β. *Then the fundamental group* $\pi_1(S^3 - \mathfrak{B})$ *of the complement of* $\mathfrak{B}$ *in* S^3 *admits the presentation:*

$$(5\text{-}2)\quad \langle z_1,\cdots,z_{2m};\ z_{2i-1} = z_{2i}^{-1},\ A_{2i-1}(z_1,\cdots,z_{2m})z_{\mu_{2i-1}}A_{2i-1}^{-1}(z_1,\cdots,z_{2m}) = A_{2i}(z_1,\cdots,z_{2m})z_{\mu_{2i}}^{-1}A_{2i}^{-1}A_{2i}^{-1}(z_1,\cdots,z_{2m}),\ i = 1,\cdots,m.\rangle$$

Moreover, every link group admits a presentation of this type, for some (non-unique) braid automorphism $\beta \in B_{2m}$.

Proof. The natural homomorphism from $B_{2m} \to M(0,2m)$ is onto, hence we may represent each element of $M(0,2m)$ by a surface mapping $f: S^2 - Q_{2m} \to S^2 - Q_{2m}$ which is the identity outside a disc D^2 which encloses the point set Q_{2m}. This mapping f then induces the braid automorphism $\beta \in B_{2m}$. Let z_i be the image of x_i under the homomorphism $i_*: \pi_1(D^2 - Q_{2m}) \to \pi_1(D^3 - \mathfrak{A})$ which is induced by inclusion. Then $\pi_1(D^3 - \mathfrak{A})$ is a free group of rank m, which admits the presentation

$$(5\text{-}3)\qquad \langle z_1,\cdots,z_{2m};\ z_{2i-1} = z_{2i}^{-1},\ i = 1,\cdots,m.\rangle$$

A straightforward application of the Van Kampen Theorem to the space $(D^3 - \mathfrak{A}) \cup_f (\tilde{D}^3 - \tilde{\mathfrak{A}})$ then yields the presentation of Corollary 5.1.1. ||

Theorem 5.1 could also have been established in another way, which illustrates that every link diagram may, in fact, be interpreted as a plat representation of a link. The number of strings will be shown to correspond to twice the number of overpasses. To establish this, we begin by noting that a link is said to have an m-*bridge representation* if it may be represented as the union of 3 subsets A, B, C of E^3 (using (x, y, z) coordinates), as illustrated in Figure 23 and defined below:

A is the disjoint union of m segments (bridges) of the line $z = 1$, $y = 0$, defined by the inequalities on the x coordinates, $1 \le x \le 1.5, 2 \le x \le 2.5, \cdots, m \le x \le m + \frac{1}{2}$.

B is the disjoint union of 2m line segments with endpoints in the planes $z = 0$ and $z = 1$, lying in the plane $y = 0$, and lying in the planes $x = 1, x = 1.5, x = 2, \cdots, x = m + \frac{1}{2}$.

C is the disjoint union of m arcs lying in the plane $z = 0$ with endpoints in the set $\{(1,0,0), (1.5,0,0), (2,0,0), \cdots, (m + \frac{1}{2}, 0,0)\}$.

An arbitrary link diagram is easily converted to a bridge representation, by converting the overpasses into the subsets $A \cup B$ defined above.

THEOREM 5.2. *If a link type is represented in an* m-*bridge presentation, then it may also be represented as a* 2m-*plat; conversely, if a link type is represented as a* 2m-*plat, then it may also be represented in an* m-*bridge presentation.*

Proof. Let $\mathfrak{B}$ be a link type, and suppose that $\mathfrak{B}$ is represented as the union of the 3 subsets A, B and C of E^3 defined above. Let $\Phi = (f_0(x,y), g_0(x,y))$ be any homeomorphism of the plane $z = 0$ which has the property that $\Phi(C)$ is a disjoint union of n line segments of the line $z = 0$, $y = 0$ defined by the inequalities on the x-coordinates $1 \le x \le 1.5$, $2 \le x \le 2.5, \cdots, m \le x \le m + \frac{1}{2}$. (Thus $\Phi(C)$ is the union of the m line segments lying directly below A.) Let $\Phi_t = (f_t(x,y), g_t(x,y)$ be an isotopy of Φ such that Φ_1 is the identity (that is, $f_1(x,y) = x$, $g_1(x,y) = y$), and $\Phi_0 = \Phi$. Define:

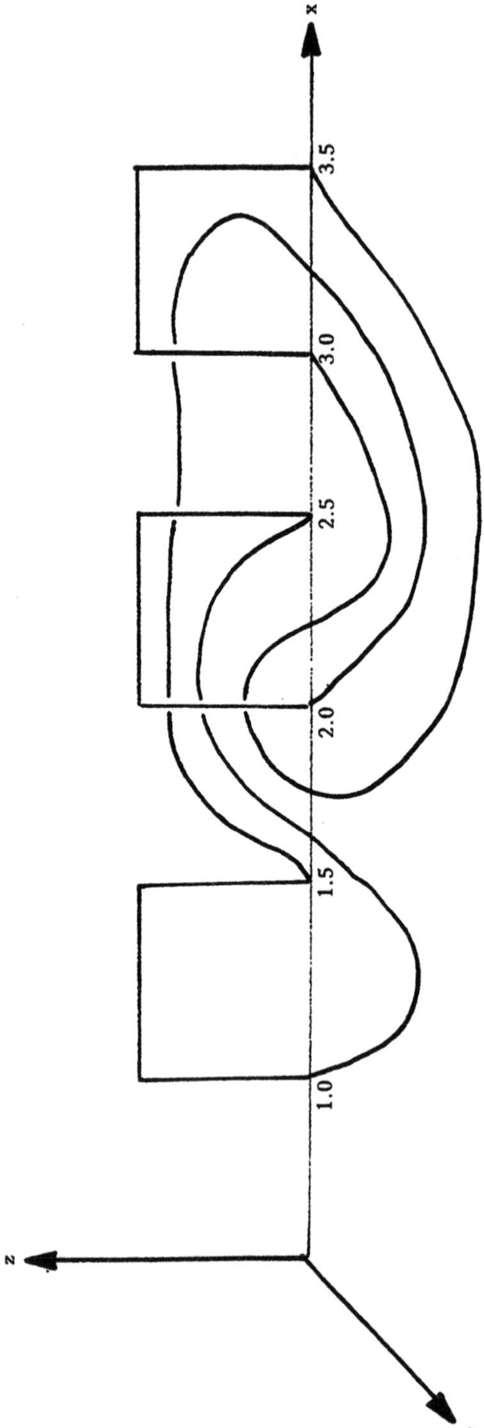

Fig. 23.

$$\begin{aligned} H(x,y,z) &= (x,y,z) \text{ if } 1 \le z, \\ &= (f_z(x,y), g_z(x,y), z) \text{ if } 0 \le z \le 1, \\ &= (f_0(x,y), g_0(x,y), z) \text{ if } z \le 0. \end{aligned}$$

Then $H(A) = A$; $H(C)$ is the union of n line segments directly below A; and $H(B)$ is the union of 2m disjoint arcs, each with one endpoint in the plane $z = 0$ and the other in the plane $z = 1$. Also, each arc intersects the plane $z = t$ in exactly one point for each $0 \le t \le 1$. Thus $H(A \cup B \cup C)$ represents $\mathfrak{B}$ as a plat. The construction may clearly be reversed to go from the 2m-plat representation $H(A \cup B \cup C)$ of $\mathfrak{B}$ to the m-bridge representation $A \cup B \cup C$. Hence Theorem 5.2 is established. ||

With the aid of Theorem 5.2, we may now establish some interesting connections between the concept of the "crookedness" of a knot type, and the array of plats which represent that knot type.

Let $\mathfrak{K}$ be a tame knot type, and let K be any representative of $\mathfrak{K}$. Let $\vec{b}$ be a unit vector, and define $\mu(K, b)$ to be the number of maxima of the function $\vec{b} \cdot \vec{r}(t)$, where $\vec{r}(t)$ is a vector parameterization of K. The *crookedness* $\mu(K)$ was defined by J. Milnor [see Milnor, 1950] to be $\mu(K) = \min_b \mu(K, \vec{b})$, and the crookedness $\mu = \mu(\mathfrak{K})$ of the knot type $\mathfrak{K}$ to be $\min_{K \in \mathfrak{K}} \mu(K)$. By definition, μ is an integer, and $\mu = 1$ only for the trivial knot type. Milnor proved that $\mu \ge 2$ for non-trivial knot types. Milnor also established that the total curvature $\kappa(\mathfrak{K})$ of K is $2\pi\mu(\mathfrak{K})$, where $\kappa(\mathfrak{K}) = \text{glb}_{K \in \mathfrak{K}}\, \kappa(\mathfrak{K})$ and where $\kappa(\mathfrak{K})$ may be defined for any polygonal representative K of $\mathfrak{K}$ to be the sum of the exterior angles of the polygon K.

The connection between crookedness and minimum plat index of a knot type $\mathfrak{K}$ is expressed in the following result:

COROLLARY 5.2.1. *The crookedness of a knot type* $\mathfrak{K}$ *is the smallest integer* m *such that* $\mathfrak{K}$ *has a* 2m *plat representative (or, equivalently, an* m-*bridge representative).*

Proof. Consider an m-bridge knot type $\mathfrak{K}$ and a representative K of $\mathfrak{K}$ such that the projection of K on the x–y plane is an m-bridge presentation. Let $p_1, \cdots, p_{2m}$ be points on K separating the bridges from the underpasses. We can easily deform K so that the z coordinate $z(p)$ of points p on K satisfies $z(p_i) = 0$, $i = 1, \cdots, 2m$, $z(p) < 0$ if p belongs to an underpass, $z(p) > 0$ if p belongs to a bridge, z has one maximum on each bridge and one minimum on each underpass. Thus $\mu(K, (0,0,1)) = m$ and $\mu(\mathfrak{K}) \leqq m$.

On the other hand suppose that K is a representative of $\mathfrak{K}$ with $\mu(K, (0,0,1)) = m$. Let $p_1, \cdots, p_m$ be the maximum points. We may assume there are exactly m minimum points $q_1, \cdots, q_m$, and that $0 < z(p) < 1$ for every $p \in K$. By a deformation moving K only in small disjoint neighborhoods of the maxima and minima we may arrange that there are small disjoint closed subarcs $\mathcal{A}_1, \cdots, \mathcal{A}_m$ and $\mathcal{B}_1, \cdots, \mathcal{B}_m$ containing $p_1, \cdots, p_m$ and $q_1, \cdots, q_m$ respectively such that $z(\mathcal{A}_i) \equiv 1$ and $z(\mathcal{B}_i) = 0$, and $z(p)$ is strictly monotone on each of the 2m open subarc components of $K - \bigcup_{i=1}^{m} \mathcal{A}_i - \bigcup_{i=1}^{m} \mathcal{B}_i = \mathcal{C}$. Thus K is seen to be a plat representative of $\mathfrak{K}$, with the collection of arcs $\mathcal{C}$ representing a geometric braid, as defined in Chapter 1. Hence $m \leq \mu(\mathfrak{K})$. ||

The result of Corollary 5.2.1 raises the natural question: how is the *braid index* of a link (defined in Section 2.4) related to its plat index? By the proof of Theorem 5.1, every closed m-braid defines a link which may also be represented as a 2m-plat. Hence plat index ≤ 2 (braid index). To see that this is the best possible answer, we give next an example of a knot which has plat index which is strictly less than its braid index.

Consider the knot 7_4 in Reidemeister's table, which may be defined by the knot diagram in Figure 24. This knot has polynomial $4 - 7t + 4t^2$. It then follows that 7_4 has braid number ≥ 3, because a closed 1-braid is always the trivial knot, while closed 2-braids have polynomials in which all coefficients are either +1 or −1 [see 3]. The crookedness $\mu(7_4)$ is, however 2. To see this, observe that the region C in Figure 24 may be altered by a scale factor to be arbitrarily "high" and "narrow."

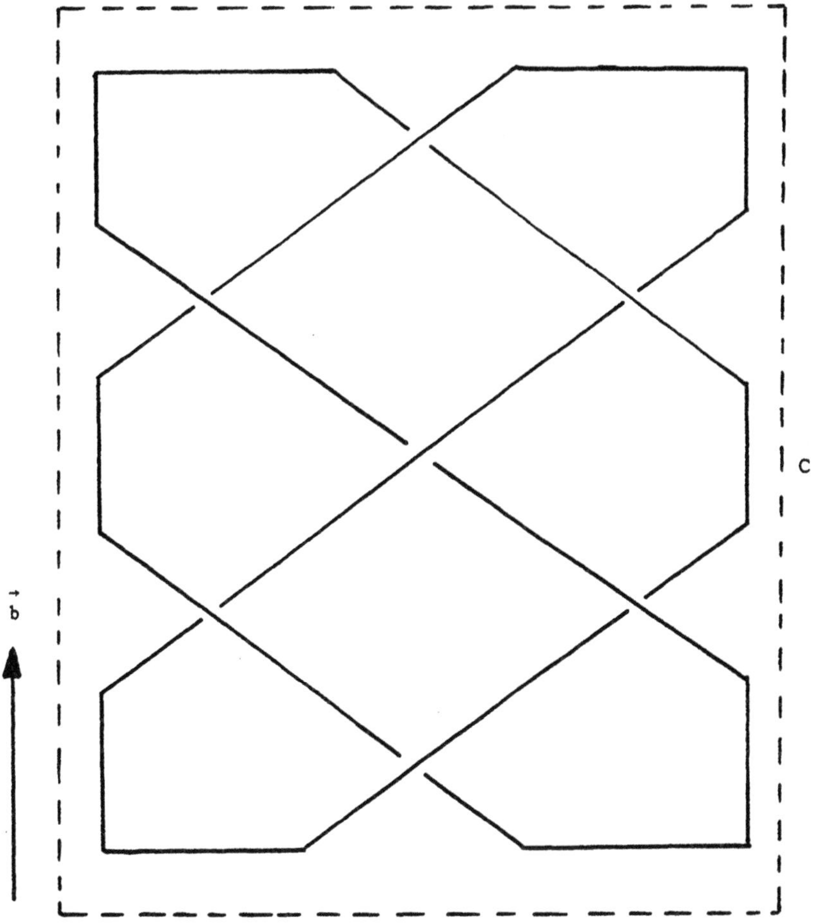

Fig. 24.

Thus we may find polygonal representatives of 7_4 which have total curvature arbitrarily close to 4π. Since 7_4 is a non-trivial knot type, its crookedness must then be precisely 2, hence its plat index is also 2.

We remark that all of these results are non-constructive. Except for special cases, we are unable to compute either the braid index or the plat index of a link type.

5.2. *The algebraic plat problem*

The association which was developed in Section 5.1 between elements of the group $M(0,2m)$ and links in S^3 raises the interesting question: when do two elements of the group $M(0,2m)$ define equivalent links? To begin to answer this question, we will need to define a certain subgroup E of $M(0,2m)$.

Each element of $M(0,2m)$ may be represented by a surface mapping f of $(S^2 - Q_{2m})$. The surface $S^2 = \partial D^3$ is a subset of the 3-ball D^3, and the point set $Q_{2m} = \partial\mathfrak{A}$ is a subset of the collection of arcs $\mathfrak{A} \subset D^3$. Let E be the subgroup of $M(0,2m)$ consisting of all mapping classes which have representatives which extend to the solid $D^3 - \mathfrak{A}$.

To describe the group E algebraically, recall that every element $\phi \in M(0,2m)$ may be lifted to an element $\beta \in B_{2m}$. In particular, if $\omega_1, \cdots, \omega_{2m-1}$ are the generators of $M(0,2m)$ used in the presentation of Theorem 4.5, then $\phi = \omega_{\mu_1}^{\epsilon_1} \cdots \omega_{\mu_r}^{\epsilon_r} \in M(0,2m)$ lifts to $\beta = \sigma_{\mu_1}^{\epsilon_1} \cdots \sigma_{\mu_r}^{\epsilon_r} \in B_{2m}$. Using the conventions in the proof of Corollary 5.1.1, the group $\pi_1(D^3 - \mathfrak{A})$ is presented by (5-3), and the kernel of the homomorphism from $F_{2m} \to \pi_1(D^3 - \mathfrak{A})$ is the normal closure N of $(x_{2i-1}x_{2i}^{-1};\ i = 1, \cdots, m)$ in F_{2m}. Hence a necessary condition for ϕ to have a representative that extends is that β leaves the subgroup N invariant. It may be shown, by the methods of [Birman and Hilden, 1974; see, in particular, Theorem 7], together with the "Handlebody theorem" [D. R. MacMillan, 1963], that this condition is also sufficient. Thus we may identify E algebraically as the image in $M(0,2m)$ of the subgroup $\tilde{E}$ of B_{2m} consisting of all braid automorphisms which leave the subgroup N of F_{2m} invariant.

THEOREM 5.3. *Let* $\phi_1, \phi_2 \in M(0,2m)$. *Let* V_1, V_2 *be the plats associated with* ϕ_1, ϕ_2. *Then* V_1 *and* V_2 *represent the same link type if* ϕ_2 *belongs to the same double coset modulo* E *as one of the following*:

$$\phi_1,\ \phi_1^{-1},\ \text{Rev}\,\phi_1,\ \text{Rev}\,\phi_1^{-1}\ ,$$

where if ϕ_1 *is defined in terms of the generators of Theorem 4.5 by the word*

(5-4) $$\phi_1 = \omega_{\mu_1}^{\epsilon_1} \omega_{\mu_2}^{\epsilon_2} \cdots \omega_{\mu_r}^{\epsilon_r} .$$

then Rev ϕ_1 *is defined by the word*

(5-5) $$\text{Rev } \phi_1 = \omega_{\mu_r}^{\epsilon_r} \cdots \omega_{\mu_2}^{\epsilon_2} \omega_{\mu_1}^{\epsilon_1} .$$

Proof. Let f_1, f_2 be homeomorphisms of $(S^2 - Q_{2m})$ which represent ϕ_1, ϕ_2 respectively. Let e_1, e_2 be homeomorphisms of $(D^3 - \mathfrak{A})$ which, when restricted to $(S^2 - Q_{2m})$ belong to the mapping classes ξ_1, ξ_2 respectively. Then we may use e_1 and e_2 to define a homeomorphism $h: (D^3, \mathfrak{A}) \cup_{f_1} (\tilde{D}^3, \tilde{\mathfrak{A}}) \to (D^3, \mathfrak{A}) \cup_{f_2} (D^3, \tilde{\mathfrak{A}})$ which gives the required equivalence between V_1 and V_2. Define h by the rule:

$$h_1 = h \text{ restricted to } D^3 = t^{-1} e_1 t$$
$$h_2 = h \text{ restricted to } D^3 = e_2^{-1} .$$

The condition that h be well-defined on $\partial D^3 = \partial \tilde{D}^3$ is

$$t f_2 h_2(p) = h_1 t f_1(p), \qquad p \in \partial D^3 ,$$

which implies and is implied by

$$f_2 = (t^{-1} h_1 t)(f_1)(h_2^{-1}) ,$$

thus

(5-6) $$\phi_2 = \xi_1 \phi_1 \xi_2 ,$$

i.e. ϕ_1 and ϕ_2 are in the same double coset modulo E.

To complete the proof, it is adequate to show that ϕ_1, ϕ_1^{-1}, Rev ϕ_1 and Rev ϕ_1^{-1} define plats which represent the same link type. This may be seen for ϕ_1 and Rev ϕ_1 by noting that the link diagram for the plat defined by Rev ϕ_1 may be obtained from that for ϕ_1 by turning the latter "upside down and over." The mappings ϕ_1^{-1} and Rev ϕ_1^{-1} determine the image of the plats determined by ϕ_1 and Rev ϕ_1 under an orientation-reversing homeomorphism of S^3. ||

Remark. Note that the results in this chapter (unlike those in Chapter 2) do *not* relate to *oriented* links. There does not appear to be any "natural" way to assign an orientation to a link which is defined by a plat.

One would certainly like to know whether the condition of Theorem 5.3 is not only sufficient, but also necessary, in order that two prime links[1] which are represented by plats on 2m strings be equivalent. We conjecture that this is true, and will prove (in Section 5.2) that it is true for $m = 2$. The question is of interest because if the conditions of Theorem 5.3 are both necessary and sufficient, then we would have a complete translation of the link problem into an algebraic problem about the group $M(0,2m)$, that is, the problem of deciding when two elements in the group $M(0,2m)$ are the same double coset modulo E. (cf. Theorem 2.3 and Corollary 2.3.1).

We note that, even if it should be true that the conditions of Theorem 5.3 are both necessary and sufficient for V_1 and V_2 to represent the same link type, this result in itself would be non-constructive unless a procedure could be found for deciding membership in a double coset modulo E in the group $M(0,2m)$. In the special case of 4-plats, treated below, we will however not only prove that the conditions of Theorem 5.3 are both necessary and sufficient, but also give a constructive procedure for applying that result.

---

1 Our conjecture is limited to prime links because one may construct counterexamples using composite links.

5.3. *The link problem for 4-plats*

In this section we will use the methods which were introduced in Sections 5.1 and 5.2 to give a constructive solution to the problem of deciding whether two given link types $\mathfrak{B}_1, \mathfrak{B}_2$ which are represented by 4-plats are equivalent. The solution to this problem will be given in Theorem 5.4.

According to Theorem 5.2, this class of links may also be described as links with 2 bridges, or alternatively as links with crookedness 2. Thus Theorem 5.4 is equivalent to an earlier result of [Schubert, 1956]. The proof which is given here is, however, new, and it is in keeping with the spirit of this text, since we will utilize the theory of braids to solve a problem about links. A proof of Theorem 5.4 which was similar to our proof was also discovered independently by O. J. Viro, who communicated his results to the author in a letter.

According to Theorem 5.3, the link problem for 2m-plats is related to the problem of finding double coset representatives for the subgroup E of the group M(0, 2m). Our approach will be to solve the latter problem for the special case $m = 2$, and then to show that in fact the solution implies a solution to the link problem. In this we will be aided by known results about the classification of 3-manifolds of Heegaard genus 1 [Brody, 1960]. We begin our study of 4-plats by investigating the structure of the group M(0, 4).

According to Theorem 4.5, the group M(0, 4) admits a presentation with generators $\omega_1, \omega_2, \omega_3$ and defining relations

(5-7) $$\omega_1\omega_2\omega_1 = \omega_2\omega_1\omega_2$$

(5-8) $$\omega_2\omega_3\omega_2 = \omega_3\omega_2\omega_3$$

(5-9) $$\omega_1\omega_3 = \omega_3\omega_1$$

(5-10) $$(\omega_1\omega_2\omega_3)^4 = 1$$

(5-11) $$\omega_1\omega_2\omega_3^2\omega_2\omega_1 = 1\ .$$

Let G be the subgroup of M(0, 4) generated by ω_1 and ω_2, and let N be the subgroup generated by $a = \omega_1 \omega_3^{-1}$ and $b = \omega_2 \omega_1 \omega_3^{-1} \omega_2^{-1}$.

LEMMA 5.4.1. *The group* M(0, 4) *is the semi-direct product of the normal subgroup* N *and the subgroup* G.

Proof of Lemma 5.4.1. That N is normal in M(0, 4) follows immediately from the following relations, which are consequences of relations (5-7)-(5-11):

(5-12) $$\omega_1 a\omega_1^{-1} = a$$

(5-13) $$\omega_2 a\omega_2^{-1} = b$$

(5-14) $$\omega_2^{-1} a\omega_2 = a b^{-1} a$$

(5-15) $$\omega_1 b\omega_1^{-1} = b a^{-1}$$

(5-16) $$\omega_1^{-1} b\omega_1 = b a$$

(5-17) $$\omega_2 b\omega_2^{-1} = b a^{-1} b$$

(5-18) $$\omega_2^{-1} b\omega_2 = a. \ \|$$

To see that M(0, 4)/N is naturally isomorphic to G, we observe that a presentation for M(0, 4)/N is obtained by adding the relation $\omega_1\omega_3^{-1} = 1$ to the defining relations (5-7)-(5-11) for M(0, 4), and that the resulting set of relations is a consequence of relation (5-7) and:

(5-19) $$(\omega_1\omega_2)^3 = 1 .$$

Thus, our result will follow if we can show that (5-19) is a true relation in G. This may be seen by noting that as a consequence of relations (5-7)-(5-9) we have:

(5-20) $$(\omega_1\omega_2\omega_3)^4 = (\omega_1\omega_2\omega_3^2\omega_2\omega_1)(\omega_2\omega_3)^3 .$$

Relations (5-10) and (5-11) then imply:

(5-21) $$(\omega_2\omega_3)^3 = 1 .$$

This, in turn, implies that (5-19) is satisfied by ω_1 and ω_2 because as a consequence of (5-7)-(5-9) we have:

(5-22) $$(\omega_3\omega_2\omega_1)(\omega_2\omega_3)(\omega_1^{-1}\omega_2^{-1}\omega_3^{-1}) = \omega_1\omega_2 . \|$$

LEMMA 5.4.2. *The group* N *is a subgroup of* E.

Proof of Lemma 5.4.2. The easiest way to see that $N \subset E$ is by pictures. Recall that we may obtain link diagrams for the link defined by a 2m-plat by regarding each ω_i as an elementary braid generator in which the ith string crosses over the $i{+}1^{st}$ (cf. Figures 22b and 22c). Suppose that $\phi \in N$, so that ϕ is a product of powers of a and b. Figure 25 shows that the maps which are associated with a and b extend to the 3-ball D^3, hence $N \subset E$. ||

We will use the symbol $\phi_1 \approx_E \phi$ to denote the fact that ϕ_1 and ϕ are in the same double coset modulo E. Using Lemmas 5.4.1 and 5.4.2, it is now clear that if $\phi \in M(0,4)$, and if we define ϕ_1 by replacing each occurence of ω_3 in the word which represents ϕ by ω_1, then $\phi_1 \approx_E \phi$.

LEMMA 5.4.3. *Let* M_2^+ *be the group of* 2×2 *matrices with integral entries and determinant* +1. *Let* $\tilde{G}$ *be the quotient group of* M_2^+ *obtained by identifying each matrix in* M_2^+ *with its negative. Then, the group* G *is isomorphic to* $\tilde{G}$ *under the mapping*

(5-23) $$\omega_1 \leftrightarrow \tilde{\omega}_1 = \begin{pmatrix} 1 & 1 \\ 0 & 1 \end{pmatrix} \qquad \omega_2 \leftrightarrow \tilde{\omega}_2 = \begin{pmatrix} 1 & 0 \\ -1 & 1 \end{pmatrix} .$$

Proof of Lemma 5.4.3. A presentation for M_2^+ may be found in Coxeter and Moser, Generators and Relations for Discrete Groups, p. 85. In terms of the generators $\tilde{\omega}_1$ and $\tilde{\omega}_2$ defined above, defining relations for M_2^+ are

The plat defined by $a = \omega_2 \omega_3^{-1}$

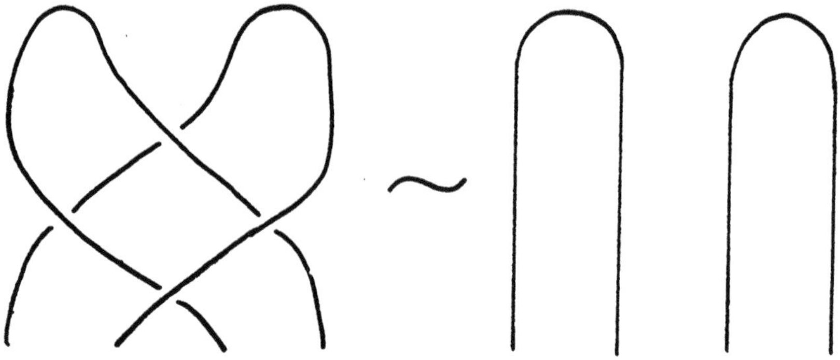

The plat defined by $b = \omega_2 \omega_1 \omega_3^{-1} \omega_2^{-1}$

Fig. 25.

(5-24) $$(\tilde{\omega}_1\tilde{\omega}_2)^6 = 1, \qquad \tilde{\omega}_1\tilde{\omega}_2\tilde{\omega}_1 = \tilde{\omega}_2\tilde{\omega}_1\tilde{\omega}_2 .$$

The element $(\tilde{\omega}_1\tilde{\omega}_2)^3$ in M_2^+ is easily computed to be the negative identity matrix, which is in the center of M_2^+. Hence, on adding the relation $(\tilde{\omega}_1\tilde{\omega}_2)^3 = 1$ to (5-24) we obtain a presentation for $\tilde{G}$, which is thus seen to be isomorphic to G under the mapping given in (5-23).

CONVENTION. In our representation of G by the matrix group $\tilde{G}$, we will adopt the convention that, if $\phi \in G$, then the matrix associated with

ϕ will be taken to be the one with a non-negative entry in the lower left corner. If that entry is zero, we will choose the matrix so that the upper left corner is non-negative.

LEMMA 5.4.4. *Let* $\phi_1 \in G$, *and suppose that the matrix representing* ϕ_1, *under the mapping defined by* (5-23), *is*

$$\bar{\phi}_1 = \begin{pmatrix} r_0 & s_0 \\ p & q_0 \end{pmatrix}, \qquad r_0 q_0 - s_0 p = 1,\ p \geq 0 .$$

Then, if $p \neq 0$, *we may find an element* $\phi_3 \in G$, *with* $\phi_3 \approx_E \phi_1$, *such that*

$$\bar{\phi}_3 = \begin{pmatrix} r & s \\ p & q \end{pmatrix} \tag{5-25}$$

where $0 \leq r, q < |p|$, *and* $r \equiv r_0 (\mathrm{mod}\ p)$, $q \equiv q_0 (\mathrm{mod}\ p)$. *If* $p = 0$, *we may choose* ϕ_3 *so that* $\bar{\phi}_3$ *is the identity matrix.*

Proof of Lemma 5.4.4. Since $\omega_1 \in E$, we may multiply ϕ_1 on the right and on the left by arbitrary powers of ω_1 without altering the double coset. The effect of right multiplication of $\bar{\phi}_1$ by $\bar{\omega}_1^k$ is to replace q_0 by $q_0 + kp$, and the effect of left multiplication by $\bar{\omega}_1^n$ is to replace r_0 by $r_0 + np$. Thus we may choose k and n, in the case where $p \neq 0$, so that $\phi_3 = \omega_1^n \phi_1 \omega_1^k$ has the sought-for properties. If $p = 0$, then $\bar{\phi}_1$ can only be a power of ω_1, and the desired form follows immediately. ||

DEFINITION. The matrix $\bar{\phi}_3$ in the statement of Lemma 5.4.4 will be referred to as a *normal matrix* associated with the element $\phi_1 \in G$.

LEMMA 5.4.5. *Let* $\phi_3 \in G$ *and* $\bar{\phi}_3 \in \bar{G}$ *be as in Lemma 5.4.4. Then, if* $p \neq 0$, *the elements* ϕ_3^{-1}, $\mathrm{Rev}\,\phi_3$, *and* $\mathrm{Rev}\,\phi_3^{-1}$ *have normal matrices:*

$$\begin{pmatrix} \bar{q} & s^* \\ p & \bar{r} \end{pmatrix}, \begin{pmatrix} q & s \\ p & r \end{pmatrix}, \text{ and } \begin{pmatrix} \bar{r} & s^* \\ p & \bar{q} \end{pmatrix} \text{ respectively,}$$

where $\bar{r} \equiv -r \pmod p$, $\bar{q} \equiv -q \pmod p$, $0 \le \bar{r}, \bar{q} < p$, *and* s^* *is the unique integer satisfying* $\bar{q}\bar{r} - ps^* = 1$. *If* $p = 0$, *then* ϕ_3, ϕ_3^{-1}, $\mathrm{Rev}\,\phi_3$ *and* $\mathrm{Rev}\,\phi_3^{-1}$ *each has the identity matrix as its normal matrix.*

Proof. If $\tilde{\phi}_3$ is given by equation (5-25), then (by the convention adopted on pp. 208-209)

$$\tilde{\phi}_3^{-1} = \begin{pmatrix} -q & s \\ p & -r \end{pmatrix}. \tag{5-26}$$

Multiplication on the left and right by appropriate powers of ω_1 then yields the stated normal form. Next we claim that

$$\mathrm{Rev}\,\tilde{\phi}_3 = \begin{pmatrix} q & s \\ p & r \end{pmatrix}, \tag{5-27}$$

which is already a normal matrix. This may be established by induction on the letter length of ϕ_3. If ϕ_3 has letter length 1, then $\phi_3 = \omega_i^\epsilon$, $i = 1$ or 2, $\epsilon = 1$ or -1. In this case $\mathrm{Rev}\,\phi_3 = \phi_3$; inspection of the matrices $\tilde{\omega}_i^\epsilon$ establishes that our assertion is true. If $\phi_3 = \omega_{\mu_1}^{\epsilon_1} \cdots \omega_{\mu_r}^{\epsilon_r}$ has letter length r, then $\mathrm{Rev}\,\phi_3 = \omega_{\mu_r}^{\epsilon_r} \cdots \omega_{\mu_1}^{\epsilon_1}$. Assume, inductively, that $\tilde{\phi}_3$ is given by equation (5-25) and that $\mathrm{Rev}\,\tilde{\phi}_3$ is given by equation (5-27). If we then increase the letter length by 1, we will replace ϕ_3 by $\phi_3\omega_i^\epsilon$ and $\mathrm{Rev}\,\phi_3$ by $\omega_i^\epsilon\,\mathrm{Rev}\,\phi_3$. The associated matrices will be

$$\tilde{\phi}_3\tilde{\omega}_1^\epsilon = \begin{pmatrix} r & \epsilon r+s \\ p & \epsilon p+q \end{pmatrix} \qquad \tilde{\omega}_1^\epsilon\,\mathrm{Rev}\,\tilde{\phi}_3 = \begin{pmatrix} \epsilon p+q & s+\epsilon r \\ p & r \end{pmatrix} \tag{5-28}$$

$$\tilde{\phi}_3\tilde{\omega}_2^\epsilon = \begin{pmatrix} r-\epsilon s & s \\ p-\epsilon q & q \end{pmatrix} \qquad \tilde{\omega}_2^\epsilon\,\mathrm{Rev}\,\tilde{\phi}_3 = \begin{pmatrix} q & s \\ p-\epsilon q & r-\epsilon s \end{pmatrix} \tag{5-29}$$

This proves that in general we may obtain a matrix representing $\mathrm{Rev}\,\phi_3$ by interchanging the diagonal entries of a matrix representative of ϕ_3. The assertion about the normal matrices of $\mathrm{Rev}\,\phi_3$ and $\mathrm{Rev}\,\phi_3^{-1}$ then follow immediately.

Before stating our main result, we summarize our results up to this point.

Algorithm for finding the normal matrices belonging to a link type $\mathfrak{B}$ *which is represented by a 4-plat.*

1. Let $\mathfrak{B}$ be a link type, and suppose that $\mathfrak{B}$ is defined by a link diagram which exhibits $\mathfrak{B}$ as a 4-plat. From the link diagram, one may recover a geometric braid on 4 strings which defines the plat, say $\beta = \sigma_{\eta_1}^{\delta_1}\sigma_{\eta_2}^{\delta_2}\cdots\sigma_{\eta_n}^{\delta_n}$. Let $\phi = \omega_{\eta_1}^{\delta_1}\omega_{\eta_2}^{\delta_2}\cdots\omega_{\eta_n}^{\delta_n}$ be the corresponding element of the mapping class group $M(0,4)$.

2. Replace each appearance of the letter $\omega_3^{\pm 1}$ in the word ϕ by $\omega_1^{\pm 1}$. This gives a new element $\phi_1 \in M(0,4) \cap G$.

3. Compute the matrix $\tilde{\phi}_1$ associated with ϕ_1 under the mapping defined by equation (5-23).

4. If the entry p in the lower left corner of $\tilde{\phi}_1$ is negative, replace each entry in the matrix by its negative. In this new matrix, if $p \neq 0$, replace the entry q_1 in the lower right corner by the unique integer q such that $q \equiv q_1 \pmod{p}$, and $0 \leq q < p$. Similarly, replace the entry r_1 in the upper right corner by the unique integer r such that $r \equiv r_1 \pmod{p}$ and $0 \leq r < p$. This determines a unique integer s such that $rq - ps = 1$. Thus we have computed the entries in the matrix

$$\tilde{\phi}_3 = \begin{pmatrix} r & s \\ p & q \end{pmatrix} \qquad rq - ps = 1, \qquad p > 0,\ 0 \leq r, q < p\,.$$

If $p = 0$, then $\tilde{\phi}_3$ may be taken to be the identity matrix.

5. A *normal matrix class* associated with a link type $\mathfrak{B}$ will be defined to be the collection of matrices

$$\left\{\begin{pmatrix} r & s \\ p & q \end{pmatrix}, \begin{pmatrix} q & s \\ p & r \end{pmatrix}, \begin{pmatrix} \overline{r} & s^* \\ p & \overline{q} \end{pmatrix}, \begin{pmatrix} \overline{q} & s^* \\ p & \overline{r} \end{pmatrix}\right\}$$

where $\overline{r} \equiv -r \pmod{p}$, $\overline{q} \equiv -q \pmod{p}$, and $0 \leq \overline{r}, \overline{q} < p$ if $p \neq 0$, or the identity matrix if $p = 0$. Note that a normal matrix class may be computed from any of its four members. The four members need not be distinct.

THEOREM 5.4. *Let* $\mathfrak{B}, \mathfrak{B}'$ *be link types which are represented by the 4-plats defined by the surface mappings* $\phi, \phi' \in M(0,4)$. *Suppose that a pair of associated normal matrices are*

$$\tilde{\phi}_3 = \begin{pmatrix} r & s \\ p & q \end{pmatrix} \quad \textit{and} \quad \tilde{\phi}_3' = \begin{pmatrix} r' & s' \\ p' & q' \end{pmatrix}$$

respectively. Then $\mathfrak{B}$ *and* $\mathfrak{B}'$ *are the same link type if and only if* $p = p'$ *and either* $q' \equiv \pm q \pmod p$, *or* $q'q \equiv \pm 1 \pmod p$.

Proof. We first establish that if ϕ and ϕ' have normal matrices which are related in the manner indicated in the statement of Theorem 5.4, then ϕ and ϕ' are of the same link type. To see this, we note first that the conditions which relate (p, q) to (p', q') are exactly equivalent to the assertion that $\tilde{\phi}_3'$ belongs to the normal matrix class of ϕ_3. Next, recall that the subgroup G of $M(0,4)$ is faithfully represented by the group $\tilde{G}$ under the mapping defined by (5-23), so that each normal matrix lifts to a unique element of $M(0,4)$. Finally, observe that the four matrices in a normal matrix class represent the same link type. Thus the condition of Theorem 5.4 is sufficient.

To prove that the condition of Theorem 5.4 is not only sufficient but also necessary, we show that if ϕ and ϕ' have normal matrices whose bottom rows are *not* related in the manner indicated in the statement of Theorem 5.4, then in fact $\mathfrak{B}$ and $\mathfrak{B}'$ are necessarily different link types. This will follow from:

LEMMA 5.4.6. *Let* $\mathfrak{B}$ *be a link type which is represented by the 4-plat* V *defined by* $\phi \in M(0,4)$, *and suppose that* $\mathfrak{B}$ *has normal matrix* $\tilde{\phi}_3$. *Then the 2-fold covering space* M_ϕ *of* S^3 *which is branched over the link* V *is one of the following: the lens space* $L(p,q)$, *or* S^3 (*if* $p = 1$), *or* $S^2 \times S^1$ (*if* $p = 0$).

If we can establish Lemma 5.4.6, then Theorem 5.4 will follow almost immediately, because if V and V' are of the same link type, then it is necessarily true that the 2-fold covering spaces of S^3 which are branched over V and V' are homeomorphic. The 3-manifolds S^3 and $S^2 \times S^1$ are inequivalent from each other and also from any of the lens spaces $L(p,q)$ because their fundamental groups are distinct.[2] By results in [Brody, 1960], the lens spaces $L(p,q)$ and $L(p',q')$ are equivalent if and only if (p,q) and (p',q') are related by the conditions in Theorem 5.4. This completes the proof of Theorem 5.4, modulo the proof of Lemma 5.4.6. ||

Proof of Lemma 5.4.6. Let X be a solid torus, which is embedded in Euclidean 3-space as illustrated in Figure 17. Thus $\partial X = T_{1,0}$. Let X' be a copy of X, which is so-related to X that a translation τ parallel to the x axis maps X onto X'. Let $\mathfrak{h}$ be an orientation-preserving homeomorphism of $T_{1,0} \to T_{1,0}$, and let $M_{\mathfrak{h}}$ be the 3-manifold which is obtained by identifying the boundaries of X and X' according to the rule $\tau\mathfrak{h}(p) = p$, $p \in T_{1,0}$. Thus $M_{\mathfrak{h}}$ is a closed, orientable 3-manifold which is represented by a "Heegaard splitting" of genus 1. [See Seifert and Threlfall, 1934 for a discussion of Heegaard splittings.]

Choose generators a, b for $\pi_1 T_{1,0}$, and generators $a' = \tau_*(a)$, $b' = \tau_*(b)$ for $\pi_1 T'_{1,0}$, with the convention that a and a' are represented by longitudes on $T_{1,0}$, $T'_{1,0}$ while b and b' are represented by meridians. Then

$$\text{(5-30)} \qquad \begin{aligned} \pi_1 T_{1,0} &= \langle a, b;\ ab = ba \rangle \\ i_* : \pi_1(T_{1,0}) \to \pi_1 X \quad &\text{by} \quad i_*(a) = a,\ i_*(b) = 1\ , \end{aligned}$$

where i_* is the homomorphism indiced by inclusion. Suppose, now, that $\mathfrak{h}_* : \pi_1 T_{1,0} \to \pi_1 T_{1,0}$ is the automorphism:

2 The fundamental group of S^3 is trivial; of $S^2 \times S^1$ is infinite cyclic; of $L(p,q)$ is cyclic of order p, where $p \neq 0, 1$.

(5-31)
$$a \to a^r b^s$$
$$b \to a^p b^q .$$

Then it may be shown without difficulty (however we omit details) that $M_{\mathcal{h}}$ is S^3 if $p = 1$, or $S^2 \times S^1$ if $p = 0$, or the lens spaces $L(p,q)$ if $p \neq 0$ or 1. The reader is referred to [Birman and Hilden, 1974] for further details on this construction.

To see the connection between our 3-manifolds $M_{\mathcal{h}}$ and the plats introduced earlier, note that the handlebodies X and X' are each invariant under a rotation i of $180°$ about the x axis in Figure 17. (This rotation also played an important role in the proof of Theorem 4.8.) Moreover, by the results of Theorem 4.6, (ii), every homeomorphism $\mathcal{h}: T_{1,0} \to T_{1,0}$ is isotopic to a product of twist maps $\mathcal{g}_{c_1}, \mathcal{g}_{c_2}$ about the curves c_1, c_2, and each of these twists may be chosen to commute with i, hence $\mathcal{h}$ may be assumed to commute with i. The orbit space of $M_{\mathcal{h}}$ under the action of i is the union of the two 3-balls D^3 and $\tilde{D}^3$, which are identified along their boundaries. Thus $M_{\mathcal{h}}/i$ is a 3-manifold which is represented by a Heegaard splitting of genus 0, hence $M_{\mathcal{h}} = S^3$. The natural map from $M_{\mathcal{h}}$ to $M_{\mathcal{h}}/i$ is a branched covering space projection, and the branch cover is the fixed point set of i, that is $(X \cup X') \cap$ (x axis). This set is a collection of four disjoint arcs (two in X and two in X') with end points in $T_{1,0} = T'_{1,0}$, which are identified via the "sewing map" $\tau\mathcal{h}$. The image of the branch cover under the covering space projection will thus be four disjoint arcs (two in D^3 and two in $\tilde{D}^3$), which are identified via the sewing map $\pi\tau\mathcal{h}\pi^{-1}$, where π is the covering space projection. Since the arcs in D^3 and $\tilde{D}^3$ are unknotted and unlinked, the branch set is easily seen to be a 4-plat (cf. Figure 22). By following through the geometry carefully (see Section 4.4), one may moreover see that if the map $\mathcal{h}: T_{1,0} \to T_{1,0}$ is expressed as a product of twist

$$\mathcal{h} = \mathcal{g}_{c_{\mu_1}}^{\epsilon_1} \cdots \mathcal{g}_{c_{\mu_r}}^{\epsilon_r}, \qquad \mu_i = 1,2;\ \epsilon_i = \pm 1$$

in the notation of Theorem 4.6, then the projection of the map h to the underlying space $D^3 \cup \tilde{D}^3$ will define the element

$$\phi = \omega_{\mu_1}^{\epsilon_1} \cdots \omega_{\mu_r}^{\epsilon_r} \in M(0,4)\,.$$

(This construction is identical to the one used in Section 4.4.) This completes the proof of Lemma 5.4.6, and hence of Theorem 5.4. ||

Example. We will now illustrate the methods of Theorem 5.4, by using it to prove that the knots 7_4 and 9_{10} in Reidemeister's knot table are inequivalent. The knot 7_4 is represented by the 4-plat $\phi = (\omega_2\omega_1^{-2})^2\omega_2$. Thus $\phi_1 = (\omega_2\omega_1^{-2})^2\omega_2$. The associated matrix is

$$\tilde{\phi}_1 = \begin{pmatrix} 11 & -8 \\ -15 & 11 \end{pmatrix}$$

and the associated normal matrix class is

$$\left\{\begin{pmatrix} 4 & 1 \\ 15 & 4 \end{pmatrix}, \begin{pmatrix} 11 & 8 \\ 15 & 11 \end{pmatrix}\right\}.$$

The knot 9_{10} is represented by the 4-plat $\psi = \omega_2\omega_1^{-1}\omega_3^{-1}\omega_2^3\omega_1^{-1}\omega_3^{-1}\omega_2$, so that $\psi_1 = \omega_2\omega_1^{-2}\omega_2^3\omega_1^{-2}\omega_2$. The associated matrix is

$$\tilde{\psi}_1 = \begin{pmatrix} 23 & -16 \\ -33 & 23 \end{pmatrix}.$$

A normal matrix is

$$\tilde{\psi}_3 = \begin{pmatrix} 10 & 3 \\ 33 & 10 \end{pmatrix}.$$

This is not in the normal matrix class of ϕ, hence 7_4 and 9_{10} are inequivalent knots.

APPENDIX: RESEARCH PROBLEMS

Many problems relating to the material in this monograph were mentioned in the course of the text. We attempt now to gather together those that seem to be of particular interest. Problems are listed in an order which corresponds (roughly) to the order in which the material was developed in the text. All are of a research nature, and many are of unknown difficulty.

1. Use the fact that each non-orientable surface has an orientable 2-sheeted covering space to carry out an analysis of braid groups on non-orientable 2-manifolds (cf. Birman and Chillingworth, 1972).

2. Let $\beta, \beta' \in B_n$ be minimal string braid representatives of link types $\mathfrak{L}, \mathfrak{L}'$ respectively. Can one find appropriate restrictions (e.g., $\hat{\beta}$ is prime, etc.) under which $\mathfrak{L} \approx \mathfrak{L}'$ if and only if β is conjugate to β or β^{-1} or Rev β or Rev β^{-1}? (Cf. Section 2.4, Example 3.)

3. Study the effect of Markov move $\mathfrak{M}_2$ on the summit power and summit form of a closed braid. Examples indicate that summit form changes drastically when $\mathfrak{M}_2$ is applied. Theorem 2.7 should play an important role in connection with this problem.

4. Conjecture: Let $\beta \in B_n$. If $\hat{\beta}$ has braid number $m < n$, then β has summit power at most 1. (The conjecture had its origins in an attempt by the author to attack problem 3 above.)

5. Study the connection between Garside's results and the notion of congruence of links (cf. remark on p. 72).

6. The conjugacy class of a braid $\beta \in B_n$ is said to be *reducible* if it has at least one representative of the form $W\sigma_{n-1}^{\pm 1}$, where W depends only on the braid generators $\sigma_1, \cdots, \sigma_{n-2}$. How can one recognize braids which belong to a reducible conjugacy class?

7. Give necessary and sufficient conditions for a link to be a positive link (cf. Theorem 2.10).

8. Find an algorithm for determining the braid index of a link. It would also be of interest to isolate special classes of links for which this decision can be made (cf. Theorem 2.8). A proof of the conjecture in problem 4 above would isolate one such class. A problem which is analogous to 8, but probably very different in content, is to find an algorithm for determining plat index of a link (cf. Corollary 5.2.1).

9. Conjecture: A knot is composite only if it can be represented by a split braid, i.e., a braid word of the form $U(\sigma_1, \cdots, \sigma_k)\, V(\sigma_{k+1}, \cdots, \sigma_{n-1})$, where the closed braids determined by U and V are both non-trivial knot types.

10. Conjecture (Murasugi): If $\hat{\alpha}$ and $\hat{\beta}$ have braid numbers n, m respectively, then $\hat{\alpha} \# \hat{\beta}$ has braid number $n+m-1$.

11. Give necessary and sufficient conditions for a link to be a pure link. Necessary conditions include:

(i) The multiplicity μ of the link equals the rank of the fundamental group of its complement.

(ii) Each component is unknotted.

Caution: Infinitely many links which satisfy (i) and (ii) above may also be constructed from "plats" defined by pure braids; see Chapter 5.

12. Is step (v) in Corollary 2.7.1 unnecessary? If not, what new words will be obtained from step (v)? A related question would be: Is there a procedure for finding the base $\overline{P}$ of a diagram of a word $D(P)$ without first computing all of $D(P)$, and is there a procedure for finding the summit tail of a word, without first computing the entire summit set $S(\beta)$?

13. Identify the kernel of the "generalized Magnus ϕ-representation of F_n," as defined in Theorem 3.8. This appears to be a problem which is capable of solution, but is difficult to organize in such a way that the solution may be generalized to arbitrary k. [See Enright, 1968.]

14. Give necessary and sufficient conditions for an $n \times n$ matrix over the ring JF_n^{ψ} to be in the Burau matrix group (cf. Theorem 3.8). This

seems to be a very difficult question. In this context, Garside's normal form for elements in B_n seems particularly useful (Theorem 2.5), because it offers the possibility of an inductive argument based on letter length.

15. Characterize, among the integral polynomials in μ variables, those which are link polynomials. This seems to be a weaker question than question 14 above. (Cf. Proposition 3.10.) This question was also posed by R. H. Fox as Problem 2 in his article "Some Problems in Knot Theory," Top of 3-manifolds, M. K. Fort, Jr., Editor, Prentice-Hall, 1962.

16. Investigate links with zero Alexander polynomial via braid theory (cf. Corollary 3.11.1).

17. Interpret Garside's solution to the word and conjugacy problems in B_n in terms of the Burau matrix representation for B_n.

18. Give sufficient conditions for a matrix in the group $\tau_B(B_n)$ to be the Burau matrix of a *positive word*.

19. If the Burau representation of B_n is faithful, one would expect that it might be a useful tool in developing a new solution to the conjugacy problem in B_n. We pose the question: is there a natural way to interpret Garside's solution to the conjugacy problem in B_n (Section 2.3) in terms of the Burau matrix group? A solution to problem 18 above would probably be a necessary first step, before one could attack this question.

20. Study the question: Can there exist words $A_1, \cdots, A_n$ in the free group F_n which satisfy:

(a) Each A_i is in the second commutator subgroup of F_n.

(b) $A_1 x_1 A_1^{-1} A_2 x_2 A_2^{-1} \cdots A_n x_n A_n^{-1}$ is freely equal to $x_1 x_2 \cdots x_n$.

A positive answer would prove that the Gassner representation is *not* faithful, and a negative answer that it *is* faithful. We conjecture (weakly) "no."

21. In the proof of Theorem 1.8 a condition was given which related to the manner in which cancellations occur in the equality (1-24) in the free group F_n. ("Either $x_\mu A_\nu^{-1}$ is absorbed by $A_{\nu+1}$, or A_ν^{-1} is absorbed by $A_{\nu+1} x_{\mu_{\nu+1}}$ for some $\nu = 1, \cdots, n-1$".) Can this condition be strengthened in any way? Can it be replaced by a stronger condition if we restrict ourselves to braids in P_n? In U_n? This question is relevant to Problem 20 above.

22. C. Miller has asked whether there is some geometric interpretation of the Burau matrices which would allow one to conclude that the representation is faithful, e.g., do they act in a natural way on some vector space which has a geometrical significance?

23. The subgroups R_{n-1} and $\ker \pi_*^k$ of Artin's pure braid group P_n are defined in the discussions preceding Theorems 3.17 and 3.18 respectively. Find a free basis for the group $\bigcap_{k=1}^{n} (\ker \pi_*^k \cap R_{n-1})$. (This might be of interest because, by Corollary 3.18.1, the group $\ker \tau_{G,n}$ is a subgroup of the aforementioned group.)

24. What are defining relations in $M(g, 0)$, if $g \geq 3$? It is possible that there is no uniform presentation for these groups for all g, so that a more reasonable question might be, what are defining relations in $M(3,0)$? This is an extremely important problem, but it also appears to be an extremely difficult problem.

25. Can one develop a purely algebraic proof of Theorem 4.6? We ask this question not because we see any particular advantage in "purely algebraic" proofs, but rather because one of the chief stumbling blocks to further progress in understanding the groups $M(g, 0)$ appears to be the fact that there is no known algebraic tool to replace the inductive arguments used by Lickorish, which are based on properties of simple closed curves on surfaces. Such a tool is needed, because it is a very difficult problem to characterize the class of elements in a surface group which have simple representatives (see Chillingworth, 1969).

26. Is every element in $M(g, 0)$ in the normalizer of an element of finite order? This question arises in connection with efforts to apply the techniques used to establish Theorems 4.7 and 4.8.

27. Are the twist generators for $M(g, 0)$ which are given in Theorem 4.6 a minimal set of twist generators for $M(g, 0)$? (Conjecture: yes.)

28. Let ζ_1 be the isotopy class of a twist about the curve c_1 in Figure 17, and let N be the normal closure of $(\zeta_1)^2$ in $M(g, 0)$. Let K be the kernel of the natural homomorphism from $M(g,0) \to M(g,0)/N$. If $g \geq 3$, is K of infinite index in $M(g,0)$? (Conjecture: yes. But if $g = 2$, K has index $6!$ in $M(2,0)$.)

29. Is the kernel of the natural homomorphism from $M(g,0)$ to $Sp(2g,Z)$ finitely generated? Finitely presented? Find a system of generators, and a system of defining relations.

30. Does $M(g,0)$ admit a faithful matrix representation of finite rank?

31. The following problem is particularly important in connection with applications of surface topology to the study of 3-manifolds: Which elements in $M(g,0)$ can be represented by maps which extend to the solid handlebody $\bar{T}_g$, and which also induce the identity automorphism on $\pi_1 \bar{T}_g$? More generally, find generators and *coset representatives* for the subgroup of $M(g,0)$ which is represented by maps which extend to the solid handlebody $\bar{T}_g$.

32. Is the condition of Theorem 5.3 both necessary and sufficient for two plats in $M(0,2m)$ to represent equivalent links, if $m \geq 3$? (For $m = 2$ the question is settled in Section 5.3.)
paper gives partial results for $m \geq 3$.)

33. Find a procedure for deciding whether two elements in the group $M(0,2m)$ belong to the same double coset modulo the subgroup E defined in Section 5.3. (This is a very difficult problem.)

34. Find an algorithm for determining the plat index of a link. This is the same as the bridge index (see Theorem 5.2), and it is also equal to the minimum number of meridian generators of the fundamental group of the complement of the link.

BIBLIOGRAPHY

1891

Hurwitz, A., "Über Riemannsche Flächen mit gegebenan Verzweigungspunkten. Math. Annalen 39, pp. 1-61.

1897

Fricke, R., and Klein, F., "Vorlesungen über die Theorie der automorphen Functionen." Vol. 1, part II, 2 (Reprinted 1965, Academic Press).

1923

Alexander, J. W., "A lemma on systems of knotted curves." Proc. Nat. Academ. Science USA, 9, pp. 93-95.

____________, "Deformations of an n-cell." Proc. Nat. Acad. Sci. USA, 9 (1923), pp. 406-407.

1925

Artin, E., "Theorie der Zopfe." Hamburg Abh. 4, pp. 47-72.

1927

Nielsen, J., "Untersuchungen zur Topologie der geschlossenen Zweiseitigen Flächen. Acta Math. 50 (1927).

1932

Burau, W., "Über Zopf invarianten." Hamburg Abh. 9, pp. 117-124.

Goeritz, L., "Die Heegaard-Diagramme des Torus." Abh. Math. Sem. Hamburg Univ. 9, pp. 187-188.

1934

Magnus, W., "Über Automorphismen von Fundamentalgruppen Berandeter Flächen." Math. Annalen. 109, pp. 617-646.

Seifert and Threlfall, "Lehrbuch der topologie." Chelsea Publishing Co., New York (reprint).

1935

Markov, A. A., "Über die freie Aquivalenz geschlossener Zöpfe." Recueil Mathematique Moscou, 1, pp. 73-78.

1936

Burau, W., "Über Zopfgruppen und gleichsinnig verdrillte Verkettunger." Abh. Math. Sem. Hanischen Univ. 11, pp. 171-178.

Fröhlich, "Über ein spezielles Transformationsproblem bei einer besonderen klasse von Zöpfen." Monatsheffe fur Math. and Physik, Bd. 44, pp. 225-237.

1938

Dehn, M., "Die Gruppe der Abbildungsklassen." Acta Math. 69 (1938), pp. 135-206.

1939

Magnus, W., "On a Theorem of Marshall Hall." Ann. of Math. 40, pp. 764-768.

Mangler, W., "Die Klassen von topologischen Abbildungen einer geschlossenen Flache auf sich." Math. Zeitsch. 44, pp. 541-554.

Weinberg, N., "Sur l'equivalence libre des tresses fermées." Comptes Rendus (Doklady) de l'Academie des Sciences de l'URSS, 23, pp. 215-216.

1942

Neuman, "On a string problem of Dirac." London Math. Soc. Jnl. 17, pp. 173-178.

1945

Markov, A. A., "Foundations of the algebraic theory of braids." Trudy mat. Inst. Steklov. 16 (Russian, with English language summary).

1947

Artin, E., "Theory of braids." Ann. of Math. 48, pp. 101-126.

———, "Braids and permutations." Ann. of Math. 48, pp. 643-649.

Bohnenblust, F., "The algebraic braid group." Ann. of Math. 48, pp. 127-136.

1948

Chow, W. L., "On the algebraic braid group." Ann. of Math. 49, pp. 654-658.

1949

Hua, L. K., and Reiner, I., "On the generators of the symplectic modular group." Trans. Amer. Math. Soc. 65 (1949), pp. 415-426.

1950

Artin, E., "The theory of braids." American Scientist, 38, pp. 112-119.

Fox, R. H., "On the total curvature of some tame knots." Ann. Math (2) 52 (1950), pp. 259-260.

Milnor, J. W., "On the total curvature of knots." Ann. of Math. 52 (1950), pp. 248-257.

1953

Fox, R. H., "Free Differential Calculus I." Annals of Math. 57, pp. 547-560.

Torres, G., "On the Alexander polynomial." Annals of Math (2) 57, pp. 57-89.

1954

Milnor, J., "Link Groups." Annals of Math. 59, No. 2, March 1954, pp. 177-195.

1956

Schubert, H., "Knoten mit zwei Brucken." Math. Zeitsch. 65, pp. 133-170.

1957

Springer, C., *Introduction to Reimann Surfaces*. Addison-Wesley Publishing Co., 1957.

1958

Fox, R. H., "Congruence classes of knots." Osaka Math. J. 10, pp. 37-41.

1959

Hu, S. T., *Homotopy Theory*. Academic Press.

1960

Brody, E. J., "The Topological Classification of the Lens Spaces." Annals of Math. 71, No. 1, pp. 163-184.

Milnor, J., "Isotopy of links," from *"Algebraic Geometry and Topology, A symposium in honor of S. Lefschetz,"* Princeton Univ. Press. Editors R. H. Fox, D. C. Spencer, etc.

1961

Clifford and Preston, "Algebraic theory of semi-groups," Vol. I. Amer. Math. Soc. Survey #7 (1961).

Gassner, B. J., "On braid groups." Abh. Math. Sem., Hamburg Univ. 25, pp. 19-22.

Klingen, H., "Charakterisierung der Siegelschen Modulgruppe durch ein endliches system definierender relatimen." Math. Ann. 144 (1961), pp. 64-82.

Lipschutz, S., "On a finite matrix representation of the braid groups." Archiv. Math. 12, pp. 7-12.

1962

Dahm, D.M., "A generalization of braid theory." Princeton Ph.D. thesis, 1962.

Fadell, E., and Neuwirth, L., "Configuration spaces." Math. Scand. 10, pp. 111-118.

Fadell, E., and Van Buskirk, J., "The braid groups of E^2 and S^2," Duke Math. Jnl. 29, No. 2, pp. 243-258.

Fadell, E., "Homotopy groups of configuration spaces and the string problem of Dirac." Duke Math. Jnl. 29, pp. 231-42.

Fox, R. H., and Neuwirth, L. P., "The braid groups." Math. Scand. 10, pp. 119-126.

Fox, R. H., Quick Trip Through Knot Theory, in *Topology of 3-Manifolds and Related Topics.* Ed. M. K. Fort, Jr., Prentice Hall.

__________, Construction of Simply-Connected 3-Manifolds, *Topology of 3-Manifolds and Related Topics.* Ed. M. K. Fort, Jr., Prentice-Hall.

Lickorish, W. B. R., "A representation of orientiable, combinatorial 3-manifolds." Annals of Math. 76 (1962), pp. 531-540.

Shepperd, J. A. H., "Braids which can be plaited with their threads tied together at an end." Proc. Royal Soc., Vol. A265, pp. 229-244.

1963

Burde, G., "Zur theorie der Zöpfe." Math. Annalen. 151, pp. 101-107.

Lipschutz, S., "Note on a paper by Shepperd on the braid group." Proc. Amer. Math. Soc. 14, pp. 225-227.

McCarty, G., "Homotopy groups." Trans. Amer. Math. Soc. 106 (1963).

MacMillan, D. R. Jr., "Homeomorphisms on a solid torus," Proc. Amer. Math. Soc. 14 (1963), pp. 386-390.

1964

Burde, G., "Über Normalisatoren der Zöpfgruppe." Abh. Math. Sem., Hamburg Univ., 27, pp. 97-115.

Coxeter and Moser, "Generators and Relations for Discrete Groups." Springer-Verlag Ergibnisse der Mathematik.

Lickorish, W. B. R., "A finite set of generators for the homeotopy group of a 2-manifold." Proc. Camb. Phil. Soc. 60 (1964), pp. 769-778.

1965

Bachmuth, S., "Automorphisms of free metabelian groups." Trans. AMS., 118, No. 6, pp. 93-104.

Smythe, N., "Isotopy invariants of links." Ph.D. thesis, Princeton Univ., 1965.

1966

Bachmuth, S., "Induced automorphisms of free groups and free metabelian groups." Trans. AMS 122, No. 1, pp. 1-17.

Lickorish, W. B. R., "A finite set of generators for the homotopy group of a 2-manifold" (corrigendum). Proc. Camb. Phil. Soc. 62 (1966), pp. 679-681.

Magnus, Karass and Solitar, "Combinatorial group theory." Interscience, division of John Wiley and Sons.

Van Buskirk, J., "Braid groups of compact 2-manifolds with elements of finite order." Trans. Amer. Math. Soc. 122, pp. 81-97.

1967

Cohen, D. E., "On the laws of a metabelian variety." J. Algebra, 5 (1967), pp. 267-280.

Magnus, W., and Peluso, A., "On knot groups." Com. Pure and App. Math. 20, pp. 749-770.

1968

Arnold, V. I., "Remarks on the branching of hyperelliptic integrals as functions of the parameter." Russian: Functional Anal. i Prilozen 2, pp. 1-3; English translation: Func. Anal. App. 2, pp. 187-9.

——————, "Fibration of algebraic functions and cohomologies of dovetails." Uspekhi Matem. Nauk. 23, No. 4, pp. 253-254.

Chein, O., "IA automorphisms of free and free metabelian groups." Com. on Pure and Applied Math. XXI, pp. 605-629.

Enright, D., "Triangular matrices over group rings." Ph.D. Thesis, New York Univ., October 1968.

Gillette, R., and Van Buskirk, J., "The word problem and its consequences for the braid groups and mapping clan groups of the 2-sphere." Trans. Amer. Math. Soc. 131, No. 2, pp. 277-296.

Quintas, L., "Homeotopy groups of surfaces." Trans. N. Y. Acad. Science, May 1968, pp. 919-938.

1969

Birman, J., "On braid groups." Com. Pure and App. Math. 22, pp. 41-72.

——————, "Mapping class groups and their relationship to braid groups." Com. Pure and App. Math. 22, pp. 213-238.

——————, "Non-conjugate braids can define isotopic knots." Com. Pure and App. Math. 22, pp. 239-242.

Chillingworth, D. R. J., "Simple closed curves on surfaces." Bull. London Math. Math. Soc. 1 (1969), pp. 310-314.

Garside, F. A., "The braid group and other groups." Quart. J. Math. Oxford 20, No. 78, pp. 235-254.

Gorin, E. A., and Lin, V. Ja., "Algebraic equations with continuous coefficients and some problems in the algebraic theory of braids." (Russia, Mat. Sbornik, Tom 78(120), No. 4; English Translation Math. USSR-Sbornik, Vol. 7, No. 4.

Gupta, C. W., "A faithful matrix representation of certain centre-by-metabelian groups." J. Aust. Math. Soc. 10, pp. 451-464.

Hudson, J. F. P., "Piecewise Linear Topology." W. A. Benjamin.

Magnus, W., and Peluso, A., "On a theorem of V. I. Arnold." Com. on Pure and App. Math., XXII, pp. 683-692.

1970

Arnold, V. I., "Topological invariants of algebraic functions II." Russian: Funkcional Anal. i Prilozen, 4, No. 2.

__________, "The cohomology ring of the group of dyed braids." Mat. Fametki 5, pp. 227-231 (Russia).

Cochran, David, "Links with zero Alexander polynomial." Ph.D. thesis, Dartmouth College, June 1970.

Cochran, D. S., and Crowell, R. H., "$H_2(G^1)$ for Tamely Embedded Graphs." Quarterly Jn. of Math., Oxford, 21, No. 81, March 1970.

Remeslennikov, V. N., and Sokolov, V. G., "Some Properties of a Magnus embedding." Algebra i Logika 9, No. 5, pp. 566-578.

Scott, G. P., "Braid groups and the group of homeomorphisms of a surface." Proc. Camb. Phil. Soc. 68, pp. 605-617.

1971

Birman, J. S., "On Siegel's modular group." Math. Ann. 191 (1971), pp. 59-68.

Brieskorn, E., "Die Fundamentalgruppe des Raumes der regulaeren Orbits einer endlichen complexen Spiegelungsgruppe." Inventiones Math. 12, pp. 57-61.

Levinson, H., "Decomposable braids." Ph.D. thesis, New York Univ., 1971.

1972

Birman, J., and Chillengworth, D. R. J., "On the homotopy group of a non-orientable surface." Proc. Camb. Phil. Soc. 71 (1972), pp. 437-448.

Birman, J., and Hilden, H., "Isotopies of homeomorphisms of Riemann surfaces and a theorem about Artin's braid group." Bull. Amer. Math. Soc., Nov. 1972.

Birman, J., "A normal form in the homotopy group of a surface of genus 2, with applications to 3-manifolds." Proc. Amer. Math. Soc., 34, No. 2, pp. 379-384.

Brieskorn, E., and Saito, K., "Artin Gruppen und Coxeter Gruppen." Inventiones Math. 17, pp. 245-271.

Goldsmith, Deborah L., "Motions of links in the 3-sphere." Ph.D. thesis, Princeton University, 1972.

Magnus, W., "Braids and Riemann Surfaces." Com. Pure and Applied Math. 25, pp. 151-161.

Murasugi, K., and Thomas, R. S. D., "Isotopic non-conjugate braids." Proc. Amer. Math. Soc., 33, pp. 137-139.

Viro, O. Ja, "Linkings, 2-sheeted branched coverings and braids." Math. USSR, Sbornik 16, No. 2, pp. 223-226 (English version).

1973

Birman, J., "Plat presentation for link groups." Comm. Pure and Applied Math., XXVI, pp. 673-678.

Birman, J., and Hilden, H., "The homeomorphism problem for S^3." Bull. AMS, 79, No. 5, pp. 1006-1009.

__________, "Isotopies of homeomorphisms of Riemann surfaces." Annals of Math., May 1973.

Cohen, David, B., "The Hurwitz Monodromy Group." Ph.D. Thesis, New York Univ.

Goldberg, C., "An exact sequence of braid groups." Math. Scand., 33, pp. 69-82.

Grossman, E., "On the residual finiteness of certain mapping class groups," to appear.

Goldsmith, D., "Homotopy of Braids: in Answer to a question of Artin," to appear in Proc. of Topology Conference held in Blacksburg, Va., May 1973.

Maclachlan, C., "On a Conjecture of Magnus on the Hurwitz Monodromy Group." Math. Z., to appear.

Maclachlan, C., and Harvey, W., "On mapping-class groups and Teichmüller spaces," to appear.

Murasugi, K., "On closed 3-braids," to appear.

Ziechang, H., "On the homotopy groups of surfaces." Math. Annalen, 206, pp. 1-21 (1973).

1974

Birman, J., "Mapping class groups of surfaces: a Survey," in Discontinuous Groups and Riemann Surfaces, Ed. Greenberg, Annals of Math Studies Series, 79, Princeton Univ. Press.

________, "Poincaré's conjecture and the homeotopy group of a closed, orientable 2-manifold," J. Aust. Math. Soc., XVII, Part 2, pp. 214-221.

Birman, J., and Hilden, H., "Heegaard splittings of branched coverings of S^3," to appear in Trans. AMS.

Goldsmith, D., "Motions of links in the 3-sphere," Bull. AMS.

Gupta, C. K., and Levin, F., "Generating Groups of Soluble Varieties." J. Aust. Math. Soc., XVII, Part 2, pp. 222-233.

Magnus, W., "Braid groups: a Survey," to appear.

Montesinos, J., "Surgery on links and double branched covers of S^3," to appear.

Thomas, R. S. D., "The structure of fundamental braids," to appear.

INDEX

LIBRARY OF CONGRESS CATALOGING IN PUBLICATION DATA

Birman, Joan S 1927-
Braids, links, and mapping class groups.

(Annals of mathematics studies; no. 82)
Bibliography: p.
Includes index.
1. Braid theory. 2. Knot theory. 3. Representations of groups.
1. Title. II. Series.
QA612.23.B57 1975 514'.224 74-2961
ISBN 0-691-08149-2

Milton Keynes UK
Ingram Content Group UK Ltd.
UKHW020845070124
435586UK00010B/490

9 780691 081496